IMAGES
of America

MOUNT WILSON
OBSERVATORY

Mount Wilson Observatory

Mountain Lake

Snow - Skiing

Sierra Madre Mountains

Trout Fishing

Horseback Riding

Pasadena

School

Shopping

Golf Course

Park

Originally an oversize postcard from a successful realtor, this image shows the convenience of living on the west end of the San Gabriel Valley and some diversions that awaited the savvy investor. Crowning it all was the Mount Wilson Observatory. (Courtesy Mount Lowe Preservation Society Inc., Michael Patris Collection.)

ON THE COVER: This undated real-photo postcard created by C.G. Bartlett features the 60-inch telescope dome following a good dusting of snow. Astronomers lived on Mount Wilson year-round and had as much as 20 inches of annual rainfall and multiple feet of snow. Once on the mountain, their daily commute to the observatories was only a short walk. (Courtesy Mount Lowe Preservation Society Inc., Michael Patris Collection.)

IMAGES
of America

MOUNT WILSON
OBSERVATORY

Michael A. Patris, Maggie Sharma, and
the Mount Lowe Preservation Society Inc.

ARCADIA
PUBLISHING

Published by Arcadia Publishing
Charleston, South Carolina

Printed in the United States of America

Library of Congress Control Number: 2023939643

For all general information, please contact Arcadia Publishing:
Telephone 843-853-2070
Fax 843-853-0044
E-mail sales@arcadiapublishing.com
For customer service and orders:
Toll-Free 1-888-313-2665

Visit us on the Internet at www.arcadiapublishing.com

Dedicated to our families:
In memory of my late husband, Dr. Om P. Sharma, and
for my daughters, Arion, Keerty, and Kavita.

For my wife, Mudd Patris, for all the years of
encouragement, support, and love.

CONTENTS

ACKNOWLEDGMENTS

This book was realized in large part due to the contributions of others. First and foremost are the many astronomers on Mount Wilson, each contributing groundbreaking advances but who do not appear in the pages of this slim volume. My personal thanks go out to the astronomy archivists and librarians at the Huntington Library, Caltech, Mount Wilson Observatory, and the Carnegie Institute. Special thanks belong to retired, now volunteer, astronomer Steve Padilla for an education in plotting sunspots. My thanks are boundless for conversations with the late Freeman Dyson, Don Nicholson, Prof. Allan Sandage, Prof. Robert Jastrow, and Thomas Cragg. My indebtedness continues to Prof. Harold McAlister, Prof. Andrew Fabian, Gilbert Clark, Bob Ecklund, Sarah (Emery) Blinn, George Elder, Anuja Navare at the Pasadena Museum of History, and others whose names are lost to me but whose tidbits of Mount Wilson lore have been invaluable. The Japanese workers who made transportation up the mountain possible have also earned everlasting gratitude. There would be no Mount Wilson Observatory without the labor of the burros, pack mules, horses, and their intrepid drivers. Very special thanks to longtime friend and photographer Steve Crise, who is also vice president and archivist of the Mount Lowe Preservation Society, for his valuable input and digitization of the image collection. Unless otherwise noted, all of the images that appear in this book are courtesy of the Mount Lowe Preservation Society Inc. (MLPSI, www.mountlowe.org). Images marked Golden West Books Collection are from the publishing company Golden West Books, founded by the late Donald Duke in 1960. This company and its holdings became part of MLPSI upon his passing in 2010. Lastly, thank you, Michael, for making the layout and images beautiful.

INTRODUCTION

Surely there never was an army which advanced like the army of science.

—Winston Churchill

From Chaldean astronomy that pictured Earth under a series of nested, rotating crystal bowls that accounted for the movements of the sun, moon, and planets to the celestial concepts of Kepler, Galileo, Newton, and Einstein, mankind has been in thrall to the mystery and majesty of the heavens. Each discovery fostered new technologies and imaginative leaps in solving the riddles of the cosmos.

In the 20th century, no institution has been more central to our grasp of the universe than the Mount Wilson Observatory. This is local history worth preserving, but it also holds a broad appeal to people everywhere who are curious about our place in the cosmos, who want to know what has been discovered and by what means, and what these findings augur for the 21st century. The strides in astronomy over the last 500 years are thrilling and, viewed from the perspective of prior times, magical.

One man's dream produced the observatory that became an institution of firsts in astronomical findings. That man was George Ellery Hale. In his lifetime, Hale established the world's four largest telescopes—the 40-inch refractor at Yerkes Observatory, still the largest refractor; the 60-inch and the 100-inch Hooker reflecting telescopes on Mount Wilson; and the 200-inch telescope at Palomar. He constructed two tower telescopes for study of the sun, a design subsequently copied throughout the world. With each he sought to address the profound questions that had vexed astronomers throughout history. How was the earth formed? How is the universe organized, where are its margins? What are stars made of? What existed before our solar system? The search employed the most powerful methods and instruments that early-20th-century science could devise. Hale devoted his life to the search, even risking his mental health in his obsession to seek answers. From the first simple coelostat to the glorious 100-inch Hooker telescope that dominated astrophysical findings for over 40 years to the extensive interferometry and adaptive optics systems of the present day, the long shadow of Hale's influence and vision on Mount Wilson is central.

A precocious, hyperenergetic child, the young Hale built steam engines and boilers, classified infusoria from the local pond, perfected magic acts, constructed an observatory in the backyard, and became enthralled by the *Iliad*, *The Odyssey*, and *Don Quixote*, poetry and prose his mother read to him when he often missed school because of ill health. This was the inspiration, he wrote, that "prepared me for a life of scientific research." In 1888, by the time he was 20 years old, Hale had confirmed the hypothesis that carbon existed on the sun and had designed a spectrohelioscope to reveal its chemical composition. The spectrohelioscope made the study of the sun's surface accessible, eventually bringing him worldwide fame. The result was a burgeoning new discipline—astrophysics.

The construction of the observatory and its surrounding structures provides an embarrassment of riches in human interest stories. The enterprise was fraught with danger, adventure, and the tapestry of successes and false starts that characterize any human endeavor. The immense contribution of the pack animals with their quirky, colorful personalities; the conundrum of carrying tonnage up a two-foot-wide trail; the near misses of funding; delays and disappointments with the massive piece of glass needed for the 100-inch Hooker telescope; and the heroism of the scientists and staff, all make for breathless reading.

Hale never tired of saying that the sun is the only star we can study in detail, and to that end, he designed a 60-foot and then a 150-foot solar telescope. Above the turbulence of atmosphere and light pollution, the steady air allowed far sharper images than the ground-based solar Snow telescope he had brought from the Yerkes Observatory at Williams Bay, Wisconsin. Scientific knowledge accumulated rapidly from the start. His almost immediate discovery of magnetic fields on the sun thrilled astronomers everywhere. Hale had discovered that magnetism exists outside of Earth.

His obsession to establish ever-larger telescopes first took the form in 1908 of a 60-inch reflecting telescope that would probe the night sky farther than any human had done before. Telescopes are time machines peering into the distant past to reveal the story of the formation of the universe. Immediately, the 60-inch reflecting telescope on Mount Wilson provided astronomers with useful spectra of very faint nebula and distant star clusters and prefigured the presence of interstellar material. As the telescope started its life on Mount Wilson, impatient Hale ordered an even larger piece of glass, a hundred inches in diameter. No one knew if such a large disk could be poured successfully. It was, and in 1917, the amazing 100-inch Hooker telescope saw first light. With both the 60-inch and the 100-inch telescopes now in use, major astronomical advances galloped. Hubble's interpretation that galaxies are racing away from us, as calculated by the relationship between the red shifted spectral lines (Hubble's Law), is perhaps the most significant. His finding that the universe is expanding changed forever the way we view our place in it, and his stunning work gave rise to the study of cosmology, how elements in the universe are organized and how they evolve.

National and international scientists flocked to the world's greatest astronomical center. From Hale to Hubble to Michelson to a mule packer who rose to become a renowned observer, the laundry list of famous Mount Wilson astronomers is a long one. The roster of illustrious repeat visitors includes such scientists as Albert Einstein and Stephen Hawking.

Hale's sense of civic duty matched his scientific pursuits. Early on, he instituted and edited the *Astrophysical Journal,* still the premier journal for astronomical research; reorganized the National Academy of Sciences; and established the International Research Council. He convinced Henry Huntington to turn his art gallery, library, and gardens into an academic center, and he transformed the small coeducational school, Throop College of Technology, into the California Institute of Technology (Caltech). In conjunction with architects, he helped design the beautiful Caltech Athenaeum, the Pasadena Public Library, and the Pasadena Convention Center. Over his lifetime, he accrued well over a dozen medals and 11 honorary degrees. If Hale, one of the most famous scientists of his day, is not a household name in the 21st century, it is largely due to his aversion to self-promotion.

Space-based telescopes have taken the place of land-based ones, but interferometry systems on land abound. In 2000, Georgia State University established on Mount Wilson the largest interferometry system of the time. Comprised of six one-meter telescopes in a baseline distance of 350 meters, the resolution is equal to the angular size of a nickel seen from a distance of 10,000 miles.

The combined findings of astronomy and physics inform how we live our daily lives. Without these discoveries we would have no GPS, no microwave ovens, no radio telescopes, no surgical gamma knives, or a host of other technologies we take for granted. Through popular books on physics, most people today have heard of black holes, the big bang theory, cosmic rays, and multiverses. A systematic look at the history of astronomy and astrophysics at Mount Wilson will go a long way in providing context for these exotic concepts. The achievements of George Ellery Hale and the Mount Wilson astronomers are nothing short of thrilling. Seeking knowledge of the jeweled dome is a quest that binds children, astronomers, and even poets. Generations of Mount Wilson scientists have plumbed the mysteries of the heavens in numbers and formulas, but it was the lure of the unknown, the awe of the glittering nighttime sky, that ignited their endeavors.

One

SACRED GEOGRAPHY

As telescopes became more powerful, the search for suitable locations to house them preoccupied early American astronomers. The place must be high enough to have stable air but not so high as to prohibit scientists from living and working on-site; it needed an adequate water supply, building materials, and civilization close enough to supply electricity and machinery. The East Coast was unsatisfactory largely due to inclement weather, and the Rocky Mountain sites lacked nearby towns. The perfect location turned out to be the southern front range of the San Gabriel Mountains in Southern California. A sweet spot of geographic perfection had been found. A rough trail to the top had been forged by Benjamin Wilson in his search for lumber for his wine and fruit industries in the valley below.

Rising just north of the city of Pasadena, the summit was first called Wilson's Peak, later Mount Wilson. A serendipitous inversion layer at 4,500 feet produced clear and stable air at the 6,000-foot peak. The area was large enough to accommodate the many buildings and telescopes that would grace its topography, and it was here that astronomy's most consequential discoveries in the first half of the 20th century occurred. "Whoever worked there believed that they were engaged in projects bigger than themselves," wrote astronomer Allan Sandage.

In 1889, Harvard University, along with the University of Southern California, established the first makeshift observatory. Director William Pickering had pronounced the site excellent for astronomical work and ordered a 13-inch telescope to be sent from Harvard, understanding that it weighed 1,600 pounds. That was far off the mark. It scaled at 3,700 pounds, impossible to transport up the flimsy trail. It came to the summit in boxes that horses pulling a wheeled dolly could manage, taking a month to complete the job. The observatory lasted only 18 months, yet during that time, considerable astronomical findings were achieved. The telescope was taken down the mountainside on skids and sent to Arequipa Peru, paving the way for Yerkes astronomer George Ellery Hale to arrive. Astronomy would never be the same.

Observatory - on Mt Low Calif.

At nearly 6,000 feet above sea level, Mount Wilson proved to be an ideal location for astronomical studies. The valley floor lay nine miles away, close enough to provide easy access once an adequate road could be built. At far left are the cottages where astronomers lived for several weeks or, in the case of permanent faculty, for several years. The "Monastery," as it was nicknamed by George Ellery Hale, housed those working on shorter projects, a comfortable if spartan series of rooms, common dining room, and a lounge with an enormous fireplace that kept the astronomers warm during harsh winter snowfalls. Though rattlesnakes and bears were frequent visitors, the cool, pine-scented air delighted all who lived on the mountain. The first astronomical permanent structure was the Snow horizontal telescope. It lies next to the 60-foot solar tower that is just to the left of the taller 150-foot solar tower. In the foreground are the domes of the 100-inch reflecting telescope and the smaller 60-inch telescope. This array of powerful eyes on the sun and the nighttime sky dominated world astronomy for over 45 years.

In 1864, Benjamin (Don Benito) Wilson, pioneer, vintner, manufacturer, politician, landowner, grandfather of World War II general George Patton, and the second mayor of Los Angeles, forged an old Indian trail to the peak of the mountain that would be named for him. With machetes and the help of Indian and Mexican workers, he painstakingly made his way up the steep, thickly overgrown Indian trail. Wilson was not searching for an astronomical site. He was seeking wood for his wine barrels and fences, but the sugar pine proved too soft for the purpose, and he abandoned the site shortly after. The trail languished for two decades, serving mainly as a refuge for horse thieves, bandits, prospectors, and hunters. Wilson did not live to see the abundance of camping sites that would spring up along his trail in the 1880s and 1890s. In his lifetime, in a challenging frontier environment, Don Benito Wilson earned an enviable reputation for integrity, inclusiveness, and incorruptibility.

Midway up the trail, Wilson built a halfway house to serve as a construction camp to extend the trail to the summit. It included a stable, blacksmith's shop, cabin, and chicken coop. Clearing a pathway through the precipitous slopes was grueling, but within six months, he had established a crude trail to the top. Halfway House later became a rest stop and shelter for hikers on their way to the summit.

Camps continued to proliferate along the trail. Hikers and sightseers filled them to capacity, lured by beauty and clear air. Orchard's Camp, formerly Don Benito Wilson's Halfway House, nestled amongst the scented trees at the head of the canyon, became especially popular. Burros were now available to take hikers the remaining four miles to the summit, costing $1 one way, $1.50 round trip.

From 1880, hikers flocked to the mountain, marveling at the profusion of wildflowers, condors, and animals in the terrain. Mountaineer A.G. Strain decided to establish a resort following an altercation with Peter Steil, a Pasadena restaurateur who had erected a thriving camp on the peak. Access to Steil's camp used the trail leading to Strain's property. Strain immediately erected a fence to stop passage, tore down tents, and sued Steil for infringing on his property rights. The court's verdict was in favor of Steil holding the trail, "a public highway that cannot be closed against travel." Strain then developed his own enormously successful camp in the grove of beautiful sugar pines near Wilson's spring, northwest of the summit. The spring provided a ready source of water to the Mount Wilson astronomical facilities, solving a critical requirement. Strain's camp thrived from 1891 until 1914, delighting throngs of sojourners who came to hike, relax in a hammock, and explore the charming botany of the mountaintop.

Above the Fog   Mt. Wilson, Calif.

There is nothing quite as magical as leaving the valley floor on a cloudy day and bursting through to instantaneous brilliant sunshine at 4,500 feet. It is this inversion layer that allows nearly 300 clear days on Mount Wilson, an astronomer's dream come true. Very few sites command this degree of viewing days at a tolerable elevation.

In 1889, William Pickering, director of the Harvard University Observatory, met with Marion M. Bovard, president of the University of Southern California, to establish a scientific center at the summit. Pickering, left, is pictured along with (in no particular order) Bovard, Judge Benjamin Eaton, lens maker Alvan Clark, and two other unidentified men making their way up the fragile trail. A doorless structure at the peak was their drafty home for the night. Pickering pronounced the site excellent.

Pickering ordered a 13-inch telescope from Harvard. Judge Eaton widened the narrowest places and smoothed out much rough terrain, forming the Mount Wilson Toll Road Company along with 18 businessmen who contributed capital to the venture. In July 1891, the new trail, now a toll road, was opened to the public at 25¢ round trip for hikers and 50¢ for riders.

Now that the telescope was on the mountain, a makeshift building was erected to house it. Though the structure was simple, the brutal winter of 1889 delayed the construction. Nevertheless, the astronomers, under the direction of Prof. E.S. King, were able to photograph over 1,150 stars within a year, mapping the nighttime sky with celestial objects never before seen.

This doughty little telescope punched above its weight. It lasted only 18 months as the noisy hikers and campers greatly interfered with the astronomer's work. However, a more compelling reason was a dispute between Harvard and the University of Southern California over land rights. A bill to grant Harvard acreage was fiercely opposed by President Bovard. The telescope was taken down the mountainside on skids and sent to Arequipa, Peru.

The second attempt to establish astronomy on Wilson's Peak took place when local entrepreneurs got into the act. Thaddeus Lowe, builder of the Great Cable Incline on Mount Lowe, proposed an electric/cog railroad to the summit where he would install a 37.5-inch refracting telescope and build a grand hotel to accommodate the swarms of visitors he envisioned. The University of Southern California had not given up hope but severe financial problems did not bode well. In the meantime, Harvard had built a 24-inch telescope with the intention of Pres. Charles W. Eliot to send it to Mount Wilson. Excited by the prospect of finally establishing a center for astronomy, local businessmen and President Eliot made their way to the top. Eliot is second from right with the glasses, Thaddeus Lowe is seated to his left, and Judge Eaton is seated at lower left. Pickering, who had formerly championed the site, was having second thoughts. He discouraged the plan so thoroughly that Eliot decided to abandon the idea, and the 24-inch telescope ended up in Arequipa, Peru, too.

The University of Southern California's plans collapsed when funding was not forthcoming. The worsening condition of Southern California economy saw an end to Lowe's wealth, but not before he had built a sizeable observatory housing a 16-inch refractor on nearby Echo Mountain in 1894, along the Alpine Division of the scenic Mount Lowe Railway. While Mount Wilson was abandoned to hikers and campers, Mount Lowe thrived with the great cable incline, a hotel, and an observatory. The Lowe Observatory operated for 32 years in all.

At 24 years of age, George Ellery Hale was awarded the directorship of the Yerkes Observatory in Wisconsin before it was even built. He had ordered a 40-inch refracting telescopic lens from Alvan Clark, still the largest refractor in the world. The long focal length, the distance between the object glass at one end of the tube and the point at which light is focused, resulted in larger images with greater detail than anything seen before. In 1897, the observatory opened. The beautiful instrument and moving parts weighed over 20 tons and was controlled by buttons on a small keyboard. But Hale envisioned telescopes triple the 40-inch diameter. Charles Yerkes, busy constructing the London Underground, had no interest in building bigger telescopes. Hale had heard of excellent conditions on Mount Wilson in far-off California and came to investigate. An exciting era was brewing.

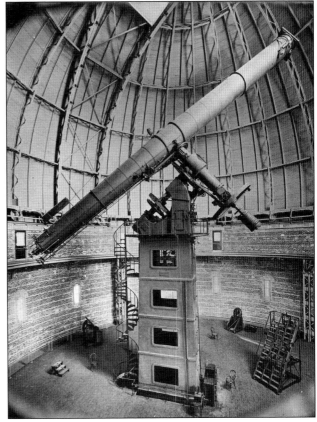

The director of astronomy, Prof. Edgar Lucien Larkin, poses in front of the Lowe Observatory. Larkin had taken over from Dr. Lewis Swift in 1900. Larkin drew in thousands of people to the site with his vivid lectures on astronomy, popularizing the subject. Though the plans of Harvard, the University of Southern California, and Lowe fell through, the way was open for George Ellery Hale of the Yerkes Observatory to establish the temple of astronomy all had sought.

In 1903, astronomer George Ellery Hale came from his lair at the Yerkes Observatory in Wisconsin to inspect conditions on Wilson's Peak. He immediately applied to the Carnegie Institution for a grant to begin studies of the sun. A total of $10,000 was awarded. A year later, Hale (No. 3) accompanied toll road president J.H. Holmes (No. 2), toll road secretary William R. Staats (No. 4), and astronomer Professor Campbell (No. 5) along with others to the summit.

Two

BARELY A TRAIL

If the main drama of the Mount Wilson story is the explosion of new astrophysical information, its subplot is transportation. Limited to foot power and mules, early astronomers struggled with the matter of how to deliver tonnage of larger and larger telescopes up an insubstantial trail.

When George Ellery Hale arrived in 1903 to scout the peak, the trail up the mountain presented a formidable challenge. Although the Mount Wilson Toll Road had smoothed and widened it a bit, the trail from Eaton Canyon was still a rough, zigzagged path with severe angles and daunting gradients. Nevertheless, Mount Wilson had become the mecca of adventurers, old and young, earning the sobriquet the "Great Hiking Era" of the late 1800s and early 20th century.

Hale believed that if the Mount Wilson site could demonstrate groundbreaking astronomical findings with a simple telescope, he would be able to raise the money needed to construct a road capable of handling the massive parts of the large telescopes he dreamed of. Plans first went forward to establish a coelostat, a 12-inch plane mirror to reflect the sun's rays through a long tube. A driving clock rotated the mirror to follow the sun. Light fell upon a lens that formed an image on a photographic plate 6.5 feet away. His dream of a center of science might be modified, but it would not be stopped.

The weight of materials was restricted to 150 pounds for burros and 200 pounds for mules. Lumber for building was limited to eight feet in length, and to this day, rounded corners can be seen on timbers that dragged a little on the ground as the equines made their way to the mountaintop. The climb up the mountain was arduous, and although the pack trains left before dawn, they often arrived at the summit after dark.

Hale's excellent findings with the Snow coelostat did indeed persuade the Carnegie Institute of Washington to fund the Mount Wilson Solar Observatory, as it was first named. The award was $150,000 for each of two years. Hale was ecstatic.

Trail up Mt. Wilson,
California.

This early postcard advertises the trail up Mount Wilson as "On the Road of a Thousand Wonders." Sightseers from the valleys below heeded the call in droves. Hikers and vacationers were instantly enchanted with the grandeur of vistas; the perfume of bay, sage, and wildflowers; the dizzying cliffs and crags; the music of waterfalls; and the crystal clear air. Invigorated, many travelers came for a rest cure, others to immerse themselves in the variety of plants and animals inhabiting a mountain so close to the city. It was a wonderland open to all and had its share of out-of-state visitors who had heard of the campgrounds. Only consumptives were not admitted. Tuberculosis was still a common and dangerous disease, and although several sanatoria existed at high elevations, Mount Wilson kept apart, reserving the mountain for science and the pleasure of sightseers.

The Old Mount Wilson Indian trail from Sierra Madre was a hazardous affair using the shortest route up vertical slopes regardless of obstacles. In 1889, it had been widened where possible for the Harvard telescope, but it was still laced with friable edges unable to withstand heavy loads. In 1891, the New Mount Wilson trail from Eaton Canyon in Altadena was widened to four feet, though it was still a rough, zigzagged path with severe angles and sharp slopes of 17 degrees and more. Now renamed the Mount Wilson Toll Road, fees were paid and burros rented at the base in Altadena.

The starting point was the Mount Wilson Toll House where fees were paid.

A rustic but welcoming Halfway House on the Mount Wilson Toll Road was built by George Schneider in a lovely grove of big cone spruce and oak. From 1897, food, soft drinks, and rooms to rent were available in a two-story structure a short distance from this first cabin.

Martin's Camp was another four miles up the trail from Schneider's Halfway House. It was a welcome sight to sojourners weary of climbing on foot or bouncing on burros. Most of the accommodations were wood-frame cottages, comfortably furnished. The excellent meals were acclaimed far and wide; hammocks dotted the landscape, nestled among the cedar, spruce, and oak groves. The summit lay another mile above, a steep climb.

The old log Casino, built by A.G. Strain, stood on the site of the dismantled Harvard Observatory. It was not a gambling casino but a pleasing site to travel-weary adventurers who gathered here for refreshments and conversation. For overnight guests, cottage tents in shaded areas dotted the landscape. Furnishings were available as required, and mail was delivered daily. A mile down the trail at Martin's Camp, telephone service to Pasadena and Los Angeles was available. Pasadena hotels prominently displayed brochures with a list of charges for both furnished and unfurnished tents at the summit. Wilson's Peak camping grounds proudly boasted repeat visitors and were filled to capacity most weekends. Overnight guests paid fees that look modest, but $1 in 1889 was the equivalent of about $33 today. Only the fairly affluent could afford an overnight stay.

Wilson's Peak Camping Privileges . . . .

One day, each person, 25c. ; week, $1 ; month, one person, $3, two persons, $5

Unfurnished Tents, 10 x 12

		WITH FLOOR	WITHOUT FLOOR
Week	1 person	$2 50	$1 75
	2 persons	4 00	2 50
Month	1 person	6 50	4 75
	2 persons	10 00	7 50

Furnished Tents, 10 x 12

		WITH FLOOR
Day	1 person	$ 75
	2 persons	1 00
Week	1 person	4 00
	2 persons	6 50
Month	1 person	12 00
	2 persons	16 00

Furnished Tents include bed, mattress, blankets, bed linen, towels, wash bowl, pitcher, mirror, table, two chairs, lamp, olla, dishes.

PASADENA AND MT. WILSON TOLL ROAD CO.,

PROPRIETORS.

This pack train is bringing supplies to the summit. The muleskinners were talented men who cajoled, motivated, and rewarded their faithful charges. They formed a close bond with the animals and treated them well. The quirky personalities of the mules and burros entertained and kept the muleskinners alert; the sometimes lurid language of the muleskinners kept the Mount Wilson staff amused.

Building specifications often had to be adjusted to meet the eight-foot lengths the pack train could carry. Nevertheless, some of the ends scraped the ground, making an already slow procession even slower. Sharp curves presented difficulty, and more than once the muleskinner held his breath as the mule stepped to the edge of the precipice in order to navigate a turn.

Month after month, in sun and snow, burro and mule trains navigated the narrow, often slippery, trail. In this 1905 photograph, food supplies and cooking equipment are making their way to the top. Supplies generally arrived in excellent shape, with one exception. One of the mules was fond of stopping on the trail and rolling over, once destroying an expensive, precious diffraction grating.

A Mountain Horseless Carriage.

In 1905, the Mount Wilson Toll Road Company built a hotel at the summit, complete with a large dining hall serving the choicest meats, vegetables, and fruits. Guests raved about the cold water, clear as crystal from nearby streams. This postcard was used as an unusual advertisement for a "Mountain Horseless Carriage" that would take even children safely by burro to the Mount Wilson Hotel.

Women took to the hiking life as much as men, happily posing for photographs on the journey up the slopes. Regardless of weather, both men and women dressed in their finest for the expedition, whether hiking or riding a burrow. Women typically rode sidesaddle, in keeping with the post-Victorian tradition.

The docile nature of the trail animals carrying passengers is apparent as an unidentified woman playfully pulls the burro's tail. The group's close proximity to the burro and the relaxed poses demonstrates the trust they placed in the behavior of the animal. The rider's heeled boots reveal that she is not about to hike on this trip and will arrive at the Mount Wilson Hotel dressed in style.

As this 1910 real-photo postcard depicts, snow did not stop the work of the pack trains, though it often slowed it considerably. The terrain was hilly with steep upgrades that became equally steep downhill grades on the way back. One of the muleskinners took his dog on round trips, much to the enjoyment of the dog.

The load is light for these lucky three carrying mail and newspapers to the summit. In those early days on Mount Wilson, first-class mail, newspapers, and the occasional magazines were delivered to the observatory daily. Often isolated for weeks at a time when working on a run, astronomers and staff eagerly awaited word from families and friends.

In order for larger and heavier parts to come to the summit, the trail had to be widened. In 1907, crews expanded it to over six feet. Though wider, it was still a harrowing road, with soft edges and hairpin turns. Postcards appeared demonstrating the new trail, cleverly depicting the very widest parts. The young muleskinner here has arranged his beasts to make the maximum impact of the widened trail.

It might look like the newly widened trail solved problems, but serious challenges persisted. To round a sharp curve with the mountain wall jutting close to the road required careful control of the animals. Backing up and inching forward to get around some of the tightest turns took all of the skill a pack train driver could muster.

Horse-drawn carriages left from some of the hotels in Pasadena to take the more affluent vacationers to the peak. It is hard to imagine that the nine-mile trip was comfortable though, with wooden wheels bumping over the uneven surface. Travel timings were strictly scheduled to avoid traffic coming in the opposite direction. (Courtesy Pasadena Museum of History.)

The first automobile to climb to the Mount Wilson Toll Road was a 1907 Franklin, driven by L.L. Whiteman in 1907. It was a thrilling trip, requiring the driver to back up and swing over to get around the sharpest curves. "We looked down a drop of a thousand feet," reported the terrified passenger, R.C. Hamlin. Nothing would induce him to make that trip a second time.

Supplies for Mt. Wilson.

Original photo by L.B.Grox Mt. Wilson.

Although the Franklin automobile and an M model, one-cylinder Cadillac managed to drive to the summit in 1907, the road was still extremely dangerous, with soft shoulders and precipices of more than a thousand feet. Citizens were discouraged from bringing automobiles up the trail and very few tried. It was up to the animals to bring food, items for the machine shop, electrical supplies, furnishings, and all smaller observatory parts to the top. Whenever possible, large parts were broken down into pieces of no more than 200–250 pounds that mules could manage. An advantage to the budget was that mules were much less expensive than automobiles. The 1914 postcard above shows the construction of the 100-inch telescope in full swing. Yet until a paved road could be constructed, the success of the observatory was largely due to the contributions of the dutiful pack animals. They deserve high praise, indeed.

Three

PERMANENT SNOW

In 1904, the Snow horizontal telescope arrived at the Mount Wilson, courtesy of Helen Snow who had originally donated it to the Yerkes Observatory. Work for the foundation and housing began. The heavier castings weighed 350 pounds and could not possibly be transported by animals. After wrestling with the problem of how to bring these pieces up the mountain, the answer came "out of the blue," as Hale joyously recorded. He imagined a narrow, steel truck that would measure 20 feet long and 20 inches wide. Low to the ground on small rubber wheels, the conveyance would be both led and followed by mules.

Other problems loomed. The light beam traveling 100 feet from the plane mirror to its 24-inch telescope needed to be shielded from ground heat that would distort the image. Hale solved the problem by setting the telescope high above the ground on massive 29-foot stone piers.

In 1905, the Snow became the first permanent astronomical structure on the peak. Here, while astronomers Fernando Ellerman, Walter Adams, and visiting astronomer E.E. Barnard daily observed solar images, Hale secured funding from the Board of Trade to construct full-scale laboratories in Pasadena and on the mountaintop to manufacture parts and optical instruments he would need for larger telescopes. The spectroscope, which measures the chemical composition of light waves, began life as a minor accessory, but in Hale's hands, it eventually became the primary instrument, with the telescope its faithful auxiliary.

The Snow coelostat was impressive. Thousands of spectral lines in the sun's atmosphere were photographed, direct photographs of the sunspots were taken, and individual gases floating above the sun's surface were visible for the first time in history. The detail was extraordinary, surpassing that of the Yerkes' 40-inch refractor. The Snow revealed to the world that sunspots are areas of reduced temperature in the solar atmosphere, one of Hale's many "firsts."

The etymology of the word "coelostat" is revealing. It means "an instrument to stabilize the heavens," a fitting description of what Mount Wilson Observatory would do with its unparalleled growth of knowledge.

Reliable pack trains transported a great deal of the construction material for the Snow telescope to the summit. This group has arrived with lumber and machine parts for the foundation of the telescope.

Before construction, ground was broken with picks and shovels. Not only was the Snow to be the first permanent solar telescope on Mount Wilson, it was also the first permanent one in the world. Until 1904, solar telescopes were portable in order to follow solar eclipses around the globe. Astronomy at that time was still largely concerned with describing a star's brightness and motion, not its origin and evolution. Mount Wilson astronomy would combine physics with astronomy to a degree never before attempted, yielding information on the physical properties of the sun and distant stars. Astrophysics was not generally a part of astronomy. George Ellery Hale changed all that.

Neither mule nor burro could carry some of the parts of the Snow that scaled in at more than 350 pounds. Hale designed this special carriage, low to the ground on small rubber wheels, that was preceded by a man leading a mule and followed by the same. The back-end mule acted as a brake, to keep the truck on the road. It was an odd contraption but ingenious.

A smaller version of Hale's fanciful carriage was pulled by a horse and steered by a worker. Another worker led the horse, and a third brought up the rear. So much of the trail was crumbly and dangerous that vigilance was required to ensure a safe journey. The workmen were careful; accidents were rare.

George Jones of Pasadena, designer and gifted stone mason, repaired and refurbished the old Casino into the first living quarters of the early astronomers, Hale, Fernando Ellerman, George Ritchey, and Walter Adams. Jones built an enormous fireplace, which was sometimes the only source of heat during heavy snowstorms. The fireplace easily accommodated logs two feet in diameter.

The astronomers needed a more permanent place to live. To that end a steady progression of burros brought hardware, bedding, and cooking supplies to the top where George Jones had constructed a comfortable dwelling nicknamed the "Monastery" by Hale. At Yerkes, Hale had found that astronomers could be controlled but their wives could not, and he vowed that his next observatory would have no accommodations for women.

Winter snowstorms were common occurrences, but January 1906 stands out from the rest when a storm of unprecedented magnitude bore down on Mount Wilson. Pack trains were stalled, power and telephone lines were down, and the Japanese workmen deserted. At over five feet, snow lay deeper and farther down the mountain than people could remember. All access to the summit was halted. The road became so soft that wheels sank to the hubs; walls that had reinforced the upslope side of the trail collapsed and would have to be rebuilt. The sack walls crumbled, and hundreds of tons of rock tumbled down, carrying the trail along. The lowest possible cost of repairs was $8,000—if repairs were not extensive. This dashed all hopes of financing a widened trail for the larger telescopes now being fabricated. The poor trail was rendered useless for three months.

Observatory, Wilson's Peak, Cal.

Because the telescope was horizontal, light traveling 100 feet from the plane mirror to its 24-inch telescope had to be shielded from ground heat. The problem was partially solved by setting the telescope 29 feet above ground on great stone piers. Sleds were built to transport the massive mountaintop stones, drawn by mules. It was exacting and dangerous work, and within a few short weeks, a worker lost his life and a mule committed suicide by plunging over a bluff, apparently unwilling to be dragooned into pulling enormous loads against staggering friction. From 1905 to 1908, the Snow telescope performed brilliantly. For it, Hale designed a spectrograph, an instrument for separating light at particular wavelengths. A clear 6.5-inch image with spectral lines fell on the instrument located in a 15-foot-deep pit. Spectral lines are akin to fingerprints, identifying each of the elements present in sunspots. The chemical composition of the sun was becoming known.

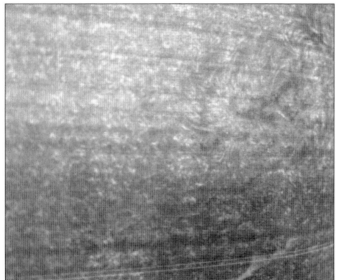

This 1908 photograph reveals the sun's surface in the light of the H alpha line of hydrogen. Swirling vortices surrounding the sunspots are visible in this very first, very indistinct capture of the phenomenon. To Hale, they resembled iron filings around a magnet, and he wondered if magnetic fields might exist on the sun. This first astonishing image caused much excitement among astronomers worldwide. (Carnegie Institute photograph; courtesy the Huntington Library, San Marino, California.)

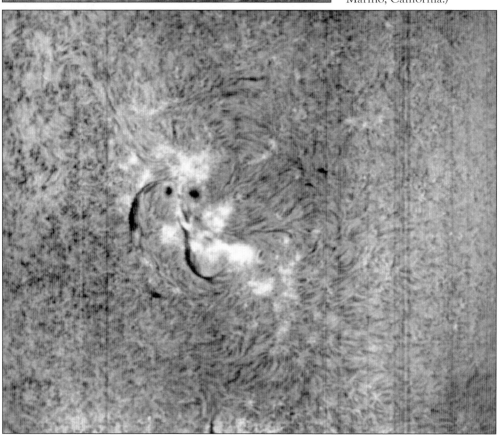

The sun is huge, a complex ball of plasma with a diameter of nearly 900,000 miles. Hale soon discovered that there are several layers to this giant gas ball, each with a different temperature and different qualities. The surface is often pockmarked with dark sunspots, sometimes many, sometimes few, and sometimes none. Astronomers were puzzled. What did it all mean?

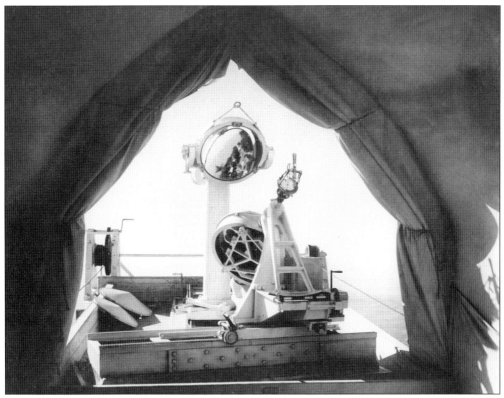

While the Snow coelostat performed admirably, it was limited. Images were distorted due to ground heat; only early morning and evening hours were useful. However, the laboratory's electric furnace that Hale had constructed analyzed the chemical elements in the lab to correspond with the sun's spectral lines. Those from sunspots matched up with cooler temperatures in the lab, the first approximation that sunspots are cooler than the disk of the sun.

During this time, the Mount Wilson Toll Road Company was busy building a hotel and 30 cottages on the summit. The hotel boasted a well-appointed dining room and excellent cuisine. The ice-cold spring water was incomparable. Finished in 1905, the hotel welcomed guests until a fire destroyed it in 1913. Another was built in its place and stood until it was razed in 1966.

Mt. Wilson Cottages.

Despite its drawbacks, the Snow delivered impressive results. Astronomers were able to photograph sunspots in fine detail. It captured the white-hot clouds of single gasses that floated above the surface and thousands of spectral lines in the sun's atmosphere for the first time in history, surpassing Hale's "greatest expectations." The scientific world paid attention.

The coelostat, a slowly revolving mirror, reflected light to a telescope's concave mirror 100 feet away. The completed building shows the long passage the beam of light must travel to produce a 6.5-inch image.

Four

TWO TOWERS

Hale found when he first scouted the peak that if he shinnied 80 feet up a tall pine, the air was cooler and more stable than at ground level. He craved greater detail of the sun's images, and an upright Snow telescope was the answer. It was an ingenious construction reaching 60 feet into the air above perturbations of the surrounding atmosphere, an invention later copied by observatories all over the world. At the top of the tower, the sun's rays are reflected from two plane mirrors through a 12-inch objective of 60-foot focal length. Completed in 1908, it was with this instrument that Hale made a great discovery: that sunspots have strong magnetic fields, the first discovery of magnetism outside of Earth. Impressive solar observations piled up, detailing the sun's corona and sunspots. Hauling tons of steel for the towers was a daunting problem. An Auto Truck carried heavy concrete loads and structural steel to the summit, but for the taller tower to come, mules again had to be used to save money.

As useful as the smaller tower was, larger images were needed in order to analyze magnetic fields deep within sunspots and to unearth the fields of very tiny sunspots. For this, greater dispersion of spectral lines was needed. To that end, Hale designed an astonishing 150-foot tower but tussled with how to make it safe from the winds that often ravaged the summit. "I'll put breeches on it!" he exclaimed, and so an inner tower was built entirely encased by a skeleton tower. By allowing a space between the two, the structures did not touch each other, providing the inner structure with its sensitive instruments a rock-solid environment.

Astronomers ascended to take sunspot measurements, at first by climbing an outside stairway and later in a bucket for two, electrically operated. The great tower rose to 100 feet above the trees, its detailed images outstripping those of the smaller tower. The Mount Wilson Observatory was becoming the center of world astronomical activity.

In order to transport structural steel and heavy equipment for the 60-foot tower, this gasoline-electric Auto Truck was specially built for the purpose by the Couple-Gear Freight-Wheel Company in Grand Rapids, Michigan. Hale had secured funding to carve out an eight-foot-wide road. Japanese laborers with picks and shovels began the laborious work of widening the nine miles to the summit.

While the Auto Truck carried heavier pieces, burros and mules were used for smaller parts, supplies, and transportation for the famous scientists who flocked to view the new instruments. Often, the hapless rider overlooked a canyon as the burro swung wide to navigate a turn. A Dr. Ridley from Sweden in an amusing effort to calm his mule, or perhaps himself, sang to him the entire way.

The tower is ready for its protective dome and a much-needed stairway. A 30-foot pit had been dug to house the long-focus spectrograph with which Hale could split spectral lines farther apart than ever in order to analyze the chemical changes at work on the sun's surface. In the lower photograph, the completed 60-foot tower stands next to the Snow horizontal telescope. Hale's exuberance to build this "high tower with no tube," as he described it, resulted in a wonderfully detailed examination of the sun's light as revealed by the spectrograph. The unknown was becoming known. With the dome and stairway complete, the astronomers took daily readings of the sun's activities in various spectra. They expected revolutionary findings. They would not be disappointed. (Below, courtesy Caltech Archives.)

These 12-inch-thick mirrors prevented warping from the rays of the sun. Light from the entrance slit was beamed to the bottom of the pit where a lens made the rays parallel before reaching the diffraction grating—a prism on steroids. The expensive grating is a reflective surface scored with thousands of parallel grooves per centimeter. Different wavelengths from the incoming light reflect at different angles. A lens focuses this diffraction producing a wide spectrum of color at the observing room, revealing which elements are present on the sun, 93 million miles away. Hale diligently searched for spectral lines that were split in two, indicating a magnetic field, and on June 25, 1908, five years to the day after his first visit to Mount Wilson, he found them. (Carnegie Institute photograph; courtesy the Huntington Library, San Marino, California.)

Peter Zeeman (1865–1943), the Dutch physicist, won a Nobel Prize in 1902 for his discovery that spectral lines will split in the presence of magnetism.

This 1908 photograph clearly depicts north and south sunspot polarity. The north (top) reveals swirls around the dark sunspot moving in a counterclockwise direction; the south (bottom) shows the swirls in a clockwise movement. It seemed a change in polarity occurred in 11-year cycles, though no explanation as to why would be forthcoming for the next 40 years. Nevertheless, these results had the attention of the world.

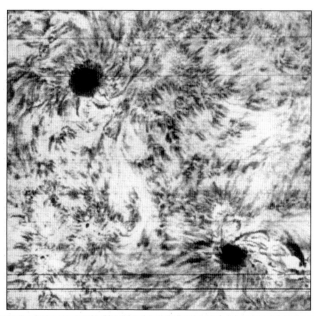

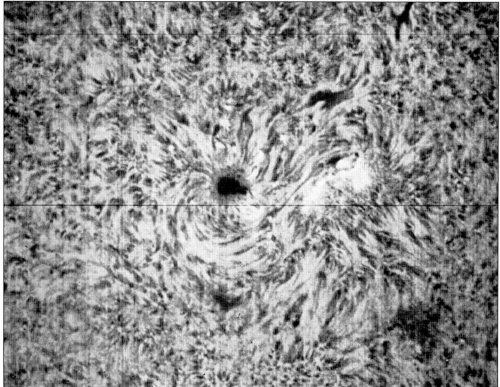

It turned out that the surface swirls Hale had thought were manifested by magnetism were actually due to events taking place in deeper layers of the sun. However, he was on the right track in finding magnetism. Tubes of interior magnetic fields are convected to the surface causing a lower temperature as they emerge. They appear dark because they are cooler. They are indeed sunspots, and they are definitely magnetic.

Living Room in Tavern,
Wilson's Peak, Cal.

The mountain was a beehive of activity. The Mount Wilson Hotel regularly hosted visitors drawn in by the advertising that flooded cities and towns in the valley. The hotel's warm tavern was a gathering spot in the evening after the day's sightseeing had concluded, but while vacationers leisurely enjoyed themselves, visionary Hale was busy designing a new, larger solar tower.

19088

This is a Receipt for Payment of Fee of 25 Cents to Enter and Visit
MT. WILSON PARK

[Subject to the Rules and Regulations of the Park Management]
On the day of its date, and will be receivable from its purchaser at the Mt. Wilson Hotel or Camps, on the day of its date stamped hereon or the following day as the equivalent of **25c** toward payment of the purchaser's hotel or camp bills for lodging.

THIS TICKET IS NOT TRANSFERABLE

There is reserved to the Manager of the Park the right to eject from the Park all persons whose presence is deemed by him undesirable.

On Honeymoon

People visited the summit for a variety of reasons. The bracing air brought health enthusiasts seeking cures of one sort or another; those with an interest in astronomy came to glimpse the towers whose findings they would read about in newspapers in the days to come. A more unusual reason is handwritten on this admittance ticket, "On honeymoon."

Hale was obsessed with the sun. By studying this star in detail, he reasoned that distant stars would be similar. When he analyzed the spectra of distant red stars, he found characteristics that matched the sunspots. He theorized that all stars could be classified by temperature, a brilliant insight. A few years hence, this would become central in understanding spectral lines according to the energy levels in the atom. Hale's fame was spreading; five of the world's most prestigious international prizes in science—the Janssen Medal, Rumford Medal, Gold Medal, Draper Medal, and Silva Medal—were already his. He would go on to accrue another nine medals. Hale was simultaneously offered the secretaryship of the Smithsonian Institution and the presidency of Massachusetts Institute of Technology. He quickly turned down the powerful positions, believing his future lay at Mount Wilson where he could achieve his dream of applying the physicists' methods to astronomical research to unravel the story of the stars. Here, he is at the spectroheliograph in the Hale Solar Laboratory in Pasadena.

The 150-foot tower goes up! An unidentified worker sits atop one of the tower's legs; the steel framing is well underway. The tower was completed in 1910, but it was not ready for use until two years later due to a failure of the lens to focus properly. This happened not once but twice. Finally, astronomers were able to use this most exacting telescope. Because of the powerful spectroheliographs, 80 feet underground in the case of the large tower, very faint light of stars could be captured in exposure times from 5 to 10 hours. The finished tower reached up to 100 feet above the trees. It held the record for the world's largest solar telescope for 50 years, until the McMath-Pierce Solar telescope was completed at Kitt Peak Observatory in Arizona in 1962.

This is the way to the dome. Astronomers went up twice daily to record sunspots after a somewhat thrilling ride in this bucket. Mount Wilson has the largest database of sunspot activity in the world. Calculator Phoebe Waterman accompanies an unnamed astronomer in the bucket. Drawings have been made every sunny day since 1917. University of California, Los Angeles solar astronomer Steve Padilla has done this work for 40 years; he now volunteers, keeping the invaluable record going.

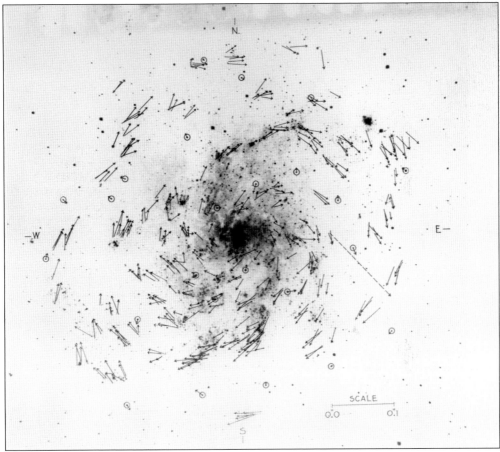

This photograph shows a sunspot drawing. Solar astronomer Steve Padilla ascends to the dome, aligns the mirrors, then travels back down to draw the 17-inch image that has formed. He takes most of the day to painstakingly mark the magnetic field strength data of every sunspot, large or small. His work is precise and beautiful. His dedication as a volunteer after funding dried up in 2014 is admirable and invaluable.

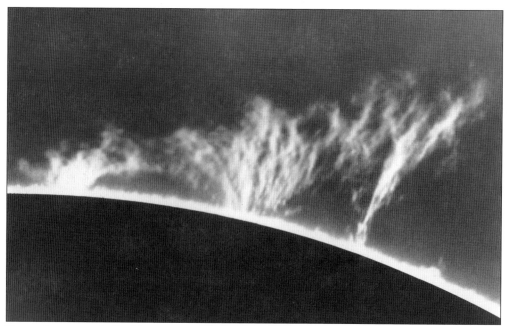

To say the sun is a ball of fire is an understatement. This roiling beast of hot gases comes in at 27 million degrees Fahrenheit at its core, cools to 10,000 degrees at the surface, and then ramps up to a whopping two or three million degrees in the corona, the sun's outermost layer. This 1908 image shows typical looping prominences 80,000 miles high, though it is now known that the largest spikes can reach over two million miles. These energetic prominences shaped by the sun's magnetic field are mad dragons that can play havoc on Earth with radio communications and the power grid. Scientists are still figuring out exactly how and why they are formed.

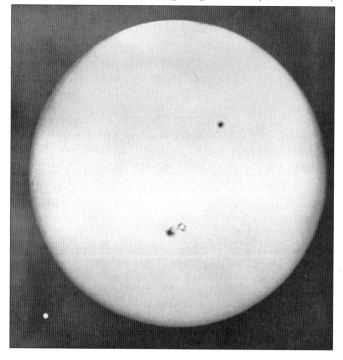

The sunspots look small on the sun but are, in fact, enormous. The white dot (lower left) represents Earth relative to the sun.

These fashionable young men, looking as louche as they can, lounge outside the Mount Wilson Hotel. The beauty of the area is now further enhanced by the science taking place on the mountaintop, drawing in a brisk business as citizens learn about the astronomer's work. Newspapers devote column upon column to the findings, making citizens both curious and proud.

This unique metal button advertises Mount Wilson with Old Man Rock situated very near the summit. The massive face seems to overlook the valley below while protecting the mountaintop from harm. It is a natural formation of stones; some say the perpetual guardian of the thousands of visitors and astronomers inhabiting Mount Wilson over the decades.

Mt. Wilson Hotel, Mt. Wilson, California.

The two completed towers stand as sentinels near the Mount Wilson Hotel on this lovely postcard. It is the epitome of serenity with the flag ruffling gently in the breeze and lofty pines, oaks, and cedars scenting the warm summer air. It is no wonder the spot was a mecca for sightseers and vacationers.

The Monastery housed resident and visiting astronomers who came to participate in the march of science taking place on the mountain. Living quarters were plain but comfortable, and a lounge with books and a fireplace offered relaxation. Meals took place in a communal dining room where resident astronomers were each given a unique napkin ring into which the napkin was replaced for the day's meals.

Five

THE ASTRONOMER'S GREED

The astronomer's greed is light. Refractors had reached their size limit with the 40 inch, but more light was needed if the stars' spectra were to be examined in the same detail as the sun's. Hale's father made yet another contribution to his son's career giving him the gift of a 60-inch optical glass disc for the mirror of a giant reflecting telescope with the understanding that Hale would have to grind it and build an observatory to house it. Even as he built the 150-foot tower, Hale hired George Ritchey, a super talented lens grinder from his Yerkes days, to start configuring the big glass at the machine shop in Pasadena. The shop was also instructed to build a driving clock, mirror support, parts of the mounting, a great steel float, and an enormous 10-foot worm gear. Ritchey would spend over two years polishing the glass "with no error in form exceeding two-millionths of an inch," a perfect parabola. The goal was to have the telescope ready at the same time the 150-foot tower was finished. It was ready, and on December 20, 1908, Ritchey photographed the Orion Nebula in stunning clarity.

The silvered, polished lens represented the lifelong quest of Hale to study the evolution of stars. His design of the telescope introduced a new engineering feat. Three focal positions, Newtonian, Cassegrain, and coudé, required one of three separate cages to be put in place, depending on the task of the astronomer that night. Only the Newtonian focus needed dark skies, so the two weeks during the dark of the moon were scheduled for those studying galactic and cosmological phenomena, "dark moon" types. The Cassegrain and coudé and their "bright moon" astronomers needed only spectroscopy, not direct viewing. This separation of dark and light moon astronomers has existed at every observatory ever since. The spectacular eye revealed stars of the 18th magnitude, millions never before seen by mankind. For the next nine years, the 60-inch telescope remained the largest time machine in the world.

Burros were still needed, even though the parts of the new telescope far outweighed those of the towers. Burros also carried visiting astronomers to the summit. One animal, with a sense of either humor or revenge, had the habit of inflating his lungs as the saddle was placed. The surprised visiting dignitary would later find the saddle slipping and sliding at some awkward point in the trail.

By the end of 1907, all 150 tons of materials for the building and the dome were on the mountain, brought up by mule teams. Only the heaviest mountings used the truck. The skeleton of the dome was riveted as welding was not yet generally practiced. The rivets added considerably to the weight of the structure, yet they convey a kind of massive grandeur that is breathtaking.

Working on the mountain was not without considerable risk from wildlife. Rattlesnakes abounded; workers were careful to wear heavy boots and keep their ears pealed for the characteristic warning rattle of a nearby snake. Moreover, antivenin medicine was not available in America until 1927 and was not in common use until the 1950s.

Bobcats roamed the summit often stalking their prey from trees. This postcard says the cat weighs 40 pounds, though how that became known remains a mystery, or at least undocumented. There was plenty of food for wild animals on the mountain, so perhaps the large size is not at all surprising.

The heaviest telescope parts were brought to the summit by the Auto Truck. The most massive piece of the mounting was the telescope tube, 6.5 feet wide and 18 feet long. The tube stands loaded at the Pasadena shop, ready for the climb. No one knew how successful the truck would be in navigating turns with such a long load. A generator operated by a gasoline engine furnished electric current to four motors placed in the wheels of the truck to negotiate sharp turns. Steep grades, however, proved too great for the load, and a team of several mules supplemented the mechanical power. The animals knew exactly when to pull with all their might and when to just trot along. The penalty for error on the still-precipitous trail was a 2,000-foot drop into the canyon.

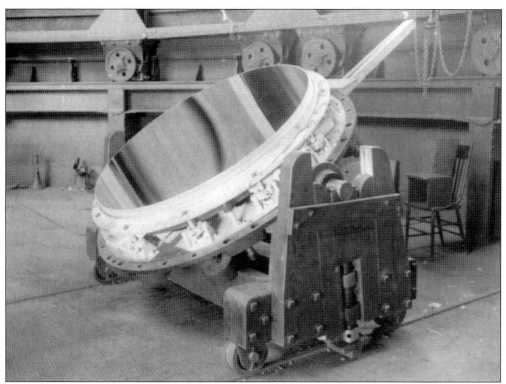

The disk of glass, 7.5 inches thick and weighing 1,900 pounds, had taken over two years to grind and polish in the dust-free lab in Pasadena. A major delay occurred when a polishing compound scratched the surface. The glass had to be reground into its perfect parabolic shape with a machine George Ritchey had designed. Finally, the mirror was perfect and was loaded onto the Auto Truck. The mood was somber as the men steeled themselves for any eventuality. A reporter from the *Los Angeles Examiner* wrote that lifting this ton of deadweight, which "the touch of a baby's hand would mar," must be the most nerve-racking of tasks. Meanwhile, the polar axle was ready to be hoisted into place. The large steel drum will float in a tank of mercury, supporting 95 percent of the telescope's weight. (Both, Carnegie Institute photograph; courtesy the Huntington Library, San Marino, California.)

The beautiful structure is up, and the dome is ready for the last step, a canvas cover that will minimize heat transfer into the observatory. A two-foot air space between the canvas and the dome acted as insulation against daytime warmth. In 1912, the canvas was replaced with a more durable metal covering. Constant temperature was maintained by neither cooling nor heating the dome but simply by being left to adjust to ambient temperature, which meant it could get pretty cold inside in winter but did not heat up too much in summer. Union Iron Works in San Francisco fabricated and tested the heaviest parts of the mounting, the lattice tube, the three cages, and the dome. All were shipped by rail to Pasadena. In total, over 200 tons of optical, electrical, and construction material had been brought to the site, much of it by mule train. It was, and is, an amazing feat. (Carnegie Institute photograph; courtesy the Huntington Library, San Marino, California.)

The telescope stands majestically inside its home, a behemoth that peers into the universe through an opening in the dome. The spacious dome rotates smoothly on rail-type wheels, two-ton gears move the telescope incrementally to compensate for Earth's rotation, and the astronomer adjusts the viewing platform up to 40 feet above the floor for the Newtonian focus. Mount Wilson had the first major telescope where light could be sent to a very large spectrograph that was not attached to the telescope, and its analysis of the spectra of stars formed the basis of modern-day astronomy. The work of the astronomer did not end with viewing. Pages of tabulation and mathematical analysis had to be carried out. Images were labeled by brightness and size, color, distance from center, exposure time, and a host of other metrics. More than 250,000 glass plates of solar and stellar images are now stored in the Carnegie-Mount Wilson office on Santa Barbara Street in Pasadena.

The telescope rests during the day, its dome closed. It remains quiet until the astronomer and night assistant set up for the coming night. If the cage needs to be swapped, a crane switched places of the one desired with the current one. The "old one" is stored on the floor of the observatory until needed again. The work is incredibly dangerous, and it was mandatory to have two people present for Newtonian and Cassegrain procedures.

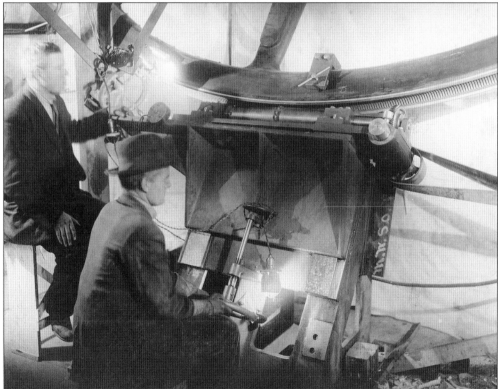

The night assistant and the astronomer worked closely to ensure that the object would appear in the center of the field by precisely computing the coordinates of the object to be viewed. Astronomers respected the night assistant who, they knew, kept a mental catalog of those astronomers who had accurately prepared and those who had not. A lot of respect and a little fear were granted to a superb night assistant.

Andrew Carnegie And Dr. Hale Beside Greatest Telescop On Earth.
Carnegie Observatory
Mt. Wilson.

In 1910, Andrew Carnegie made his only visit to Mount Wilson. Originally a poor Scottish immigrant, Carnegie (accent on the second syllable, not the first) made his wealth in steel. Impressed by Hale's results with solar telescopes, he became an enthusiastic supporter of expanding astronomy by funding large stellar telescopes on the mountain. Carnegie's interest in astronomy, luckily fueled by Hale's vision to accomplish great astronomical advances, fitted nicely with Carnegie's concept to establish research in fields not covered by existing universities. His technique was to identify the exceptional person with vision and permit him to work freely. Awed and enchanted with Hale's enthusiasm and brilliance, he financially backed the Mount Wilson Observatory and later partially the Mount Wilson and Palomar Observatories until 1979, when the Carnegie Institution moved its interest in astronomy to the southern hemisphere. Carnegie died in 1919, never having the chance to see the new 100-inch telescope on Mount Wilson. Nevertheless, he found his sole visit to the 60-inch telescope immensely gratifying, as he witnessed firsthand the dazzling splendor of the universe.

Carnegie's wife and daughter had accompanied him to Pasadena. They joined Hale's wife and daughter in a visit to the machine shop in Pasadena. Pictured here from left are Louise (Whitfield) Carnegie, Margaret Carnegie, Evelina (Conklin) Hale, Margaret Hale, and Andrew Carnegie. (Courtesy Caltech Archives.)

Both Andrew Carnegie and George Hale wanted the public to have access to the wonders of the sky through the grand telescope. Generally, one evening a week was reserved for public viewing. The telescope generated a vast amount of interest in the valley below; people arrived in large numbers to the observatory and were both dazzled and educated, viewing perhaps Saturn and an immense globular cluster on the same night. (Courtesy Caltech Archives.)

The stunning photograph above was the first exposure made with the telescope. Ritchey photographed the Great Nebula in Orion on December 24, 1908, two weeks after the telescope was completed. The nebula is visible from Earth as one of the brightest in the sky. Nebulae are gargantuan areas of dust and gas, tens to hundreds of light-years across. Keep in mind that one light-year is nearly six trillion miles! It is estimated that the Milky Way contains some 20,000 nebulae, but only about 3,000 have been catalogued. Dust obscures most. The telescope became the workhorse of the observatory. Comets, globular clusters, double spiral nebulae, and comparison of star spectra with that of the sun all began to yield their secrets to the astronomers. At left is a magnificent photograph of the Sagittarius Globular Cluster with an exposure time of 3.5 hours. (Both, Carnegie Institute photograph; courtesy the Huntington Library, San Marino, California.)

From 1912 on, citizens could drive to the summit, still a very scary trip. Nevertheless, the intrepid Mount Wilson stage left twice a day from a Pasadena office. The company did a thriving business as even that bumpy, harrowing ride seemed luxurious compared with the slowness and discomfort of a burro.

This 1913 Ford Model T prepares to take two women to the summit. They are dressed for windy conditions, although great speed would not be a hallmark of their excursion. The trip was daunting but also thrilling, giving a real sense of courage to the adventurous. Women overwhelmingly chose the auto over burro, though many men saw laborious ascent on the back of an animal a badge of honor.

Exposure time is everything when looking for great detail in a slice of the nighttime sky. Exposure time for this 1910 photograph of Ursa Major, Spiral Nebula, was four hours and 15 minutes. The spiral formation is clearly shown, and a few bright stars shine through the dust and gas.

In contrast, the same nebula photographed for seven hours and 30 minutes reveals many more stars in both the central portion and in the arms of the spiral. To obtain useful images from very distant stars, an astronomer would sometimes need to take exposures over a period of four nights. The Big Dipper in Ursa Major is comprised of the seven brightest stars in the nebula.

The serenity of this Mount Wilson Hotel cottage sheltered among the glorious pines belies the fervid activity on the mountaintop. A large workforce grew as well as the number of staff astronomers. Hale avidly fostered international relations in the world of science, inviting astronomers from all over the globe to work on projects of their particular interest. Both visitors and residents found refreshment from their labors in the variety of plant and animal life that proliferated on the mountain. Deer visited and were tamed. Some residents kept pet dogs or cats. Enormous yuccas unfailingly delighted all.

Snow confined astronomers to cabins where they played pool, wrote papers, and traded stories. Volumes could be written about the near mishaps of managing 22 tons of moving equipment to track a star, the unpredictable burros, or a mouse that stole food.

This photograph of Andromeda, Great Nebula, taken on October 13, 1909, reveals a large swirling mass of gas and dust in a thickly populated nest of stars. Many exposures were taken of Andromeda with the powerful 60-inch reflector. Longer exposures disclosed more and more detail, raising questions as to its size and placement in the sky. Was it part of the Milky Way? Several prominent astronomers believed it was as the prevailing theory held that the Milky Way is the only galaxy that exists, or could it be a separate galaxy? Other astronomers had a hunch that it might just be true that other galaxies exist far outside of the Milky Way. That was the burning issue of the day as astronomers eagerly studied nebula after nebula. The answer would not come for another 16 years, but when it did, it forever changed the understanding of the universe.

Six

THE LIGHTBRINGER

Before the 60-inch reflector was finished, impatient Hale began constructing an even larger telescope with a lens 100 inches in diameter. Could any firm cast such a huge single piece of glass? Would it sag under its own weight? Hale gambled, and with start-up money provided by business entrepreneur John D. Hooker of Los Angeles, he ordered Saint-Gobain Glassworks in France to begin the casting. The glass arrived in Pasadena but was declared unusable. The second broke in the annealing furnace; the third was too thin. Hale and Adams revisited the first one and decided it might work after all. Grinding the mighty piece began. Ritchey was not happy, declaring it would never be right. Nevertheless, Hale insisted and the work went on, but not before Ritchey spread seeds of discord to Hooker. Hale now needed to find another source of money to finance the dome, at the cost of a half a million dollars. Fortunately, Andrew Carnegie, charmed by the output of the 60-inch telescope, supported the venture.

Massive equipment arrived at the summit over a road that was still precarious. During the construction, 650 tons of material were brought to the top. First, a great pier 33 feet high was built to support the mount for the telescope. Steel supports for the dome rose. Inside the dome, the walls were lined with cork to absorb moisture. A gigantic clock mechanism was constructed to regulate the speed of the telescope as it traced the arcs of the stars. The tube, moved by the force of a two-ton falling weight transmitted to the drive gear on the polar axis of the telescope, worked like a precise Swiss watch, except it was 17 feet across! The mercury flotation system perfected the smooth movement of the telescope. Interchangeable observing cages on the upper section were used for different projects. Transportation danger remained. Riding in a truck carrying cement, astronomer Walter Adams almost lost his life when the truck toppled over the edge landing 300 feet below, taking the driver with it. Miraculously, no one was hurt.

The base of the telescope is underway. The wooden framing is in place, and the concrete siding is being added. The circular floor at the top will be the 54-foot diameter observing floor, a solid foundation for the telescope held well above ground. The pier was completed two years before the dome, and all of the materials were brought to the summit by the faithful mule trains.

The pier on which the huge telescope dome will sit must be isolated from the rest of the structure in order to prevent vibrations from wind or the movement of the dome. The pier stands 33 feet high, extending out at the top to the circular observing floor. There are four floors within the hollow pier for storage, silvering equipment, and for cooling the mirror.

The dome was built by the Morava Construction Company in Chicago, Illinois. Hale requested that it be tested on-site before being shipped. It was imperative that all parts fit together perfectly before coming to Mount Wilson. The dome was assembled in an empty lot, dismantled, and then shipped by rail to Pasadena. Early on, Henry Huntington had generously offered to transport by rail, free of charge, anything Mount Wilson needed. (Carnegie Institute photograph; courtesy the Huntington Library, San Marino, California.)

The dome is reassembled on Mount Wilson secure in the knowledge that it would function perfectly.

Critical to the success of the telescope is the smooth rotation of the dome. This photograph shows a 50-foot arm radiating from the center of the building with a motor-driven grinding machine attached. Thus, wheels on the dome were assured of faultless movement on the super-smooth rails.

Many small wheels allow the massive dome to glide flawlessly on its track giving the astronomer freedom to point the telescope to the area of the sky of interest that night. It is a feat of very beautiful engineering, making possible easeful movement of the 600-ton structure.

Mack trucks had transported 1,800 tons of various parts and materials to the summit, but something different was needed for the 11-ton pedestal pieces. "Capacity 13,000 pounds. Do not overload," read the sign on the 6.5-ton Saurer truck. A Mr. Stoner of the Mack Motor Truck Company of Los Angeles decided to make a practice run by taking the truck to the summit with an eight-ton piece, a ton and a half over capacity. There were no major problems, and so he decided to take a chance on the 11-ton mount making the journey. The huge piece was placed on the low-slung chassis. A Mack truck ahead carried three tons of material to be used for towing or holding the Saurer to the road by use of chains; another Mack followed, loaded with timber to support the edges of the road. Companies were quick to take advantage of the extraordinary event about to take place. Firestone tires on a Saurer truck, as seen in this advertisement, performed admirably. No doubt the widespread publicity boosted sales.

The Saurer truck is loaded in Pasadena. The 13-ton load is firmly anchored to the truck with chains and ropes, adding significantly to the weight. Grades averaged 12 percent; a few rose to 18 percent, throwing the center of gravity far back. Several men walked along, ready to pile on the radiator and front axle to keep the front wheels on the ground.

A signalman was crucial to the endeavor, as the position of the large beam was such that Mr. Stoner could not see the road to his left and would have to rely solely on the signals wigwagged to him by the signalman. A Mr. Smith of the Mack Motor Truck Company ably filled that post.

There was no reprieve, and nerves were stretched taut. Every bit of the trail presented difficulties; it was one continuous route of twists and turns. Backing and filling was no easy matter with the long load and heavy weight, yet it was often the only way to negotiate a hairpin turn. Mr. Stoner reported that one mile felt like 10.

The trail was fraught with danger both for the steel girder and the men surrounding it. Here, a large boulder, weighing 1,200 pounds, fell from the side wall. It struck the girder, but miraculously did not cause damage. The road had to be cleared of smaller debris that came down with the boulder, delaying the journey.

The Saurer truck had left Pasadena at 8:45 a.m. on December 12, 1915. Chances were largely in favor of failure, but the amazing vehicle and its cargo arrived safely at the dome six hours and 25 minutes later. The help of the lead Mack truck was not needed, and the Saurer's radiator ran cool the entire way. The truck stayed largely in first gear, stopping only for a 30-minute lunch break for the men. An escort of 50 men walked along with the truck, and star cameramen were assigned by two motion picture companies covering the event. The feat was monumental in the annals of astronomy; at any point, the penalty for inattention could be a sheer drop of 2,000 feet.

The matching girder was brought up without incident the following day; workmen began placing the girders in the closed-yoke mount.

Things did not go as smoothly for the 11-foot diameter tube assembly. A short distance from the summit, the road collapsed beneath the Mack truck. Rains had softened the edge of the road, and a back wheel sank into the shoulder. With block and tackle, manpower, and mules and ropes, the wheel was lifted and the shoulder reinforced with boards. The rescue took five hours.

The first and heaviest part of the telescope tube assembly is in place. Several more sections will be added, but, at least for now, the crew felt confident that the parts would reach the summit without mishap. The two round mercury floats at either end supported 95 percent of the telescope's weight.

The tube assembly grows. A second cage is installed, and another one waits in the wings. A crane is rigidly attached, traveling along the curvature of the dome. Cages are interchangeable by this method, depending on the focus of the astronomer's work. Placing was an exacting and dangerous business. A slight miscalculation could end in disaster for both man and telescope. There were a few near mishaps.

When the tube was assembled, the telescope peeked out through the open shutter of the dome. OSHA (Occupational Safety and Health Administration) did not exist; the men walked around the tube without benefit of safety belts. A worker (usually the night assistant) would step from the Newtonian platform into space 40 feet above the floor to land on a catwalk attached to the tube and then reverse the procedure once the desired cage was bolted in place. It was 45 minutes of intense concentration.

The dome is ready for the mirror.

George Ritchey designed a grinding machine to configure the 9,000-pound piece of glass into the telescope's mirror. It sits here in a horizontal position; the grinding tool is above and to the right. The mirror emerged as a perfect parabola with no error in form exceeding two millionths of an inch. It had taken over five years to get it just right. There was a cost. (Carnegie Institute photograph; courtesy the Huntington Library, San Marino, California.)

Ritchey became possessive of the telescope and began insisting on redesigning it. A falling out between Hale and Ritchey led to his dismissal as project manager. From 1912 on, his only job was to finish grinding and polishing the glass. Ritchey was a gifted observer, producing fine images of stars early on. Plates were understood to be the property of the observatory, but Ritchey removed his when he left, and they have never been found.

The dome is waiting for its powerful heart. The mirror is the means by which the astronomical work is achieved, but getting the delicate mirror up the trail without misfortune was no easy task. A boulder cascading down from the upslope would end dreams. The giant round stood 14 feet above ground when it was loaded onto the Mack truck. Two hundred men walked beside it, ready to help with ropes and boards whenever needed. The structure and size of the Milky Way galaxy, the stars, and their life stages were puzzles that could now be addressed, maybe even solved.

The Saurer truck was a godsend. On July 17, 1917, with the First World War still raging, the mirror was successfully delivered to the summit, to the immense cheers of astronomers and staff eagerly awaiting its arrival.

One of the final installations was the massive telescope drive mechanism comprised of hundreds of pounds of steel. The precision clock drive pointed the telescope accurately to the star or planet being studied, compensating for Earth's rotation to keep the object in sight. Moreover, direct current motors, counter weights, ebonite panels, and more were necessary for the very complicated electrical system of the dome and telescope. (Carnegie Institute photograph; courtesy the Huntington Library, San Marino, California.)

This is a stunning photograph of precision instrumentation. A brass and steel clock drive in motion is a captivating sight to behold. The gentle hum of working parts, the darkness of the night, and the gentle movement of the dome as the telescope tracked its target star are memories astronomers cherished and often wrote about.

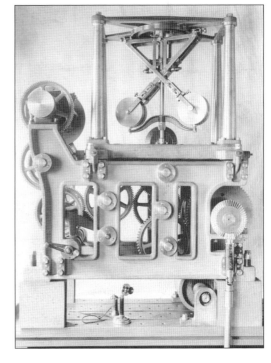

The telescope stands in all its glory. For an instrument weighing 100 tons, it is remarkably quiet and peaceful, even when in motion. Hale's vision and the brilliant work of many gifted scientists and workmen culminated in one of the most significant instruments to fathom the universe. On the lower left is the bentwood chair Edwin Hubble sat on as he peered into the starry sky at the Newtonian focus, a secondary mirror that took over seven months to shape to the required flat figure on a two-by-three-foot surface. Earlier, the optical accuracy of the majestic mirror had been confirmed when the light reflected from it onto a photographic plate more than 42 feet away focused within six thousandths of an inch of the plate's surface, a perfect parabola. This colossal telescope, whose findings would be interpreted by gifted astronomers, blazed the path to address riddles of transcendental importance to mankind.

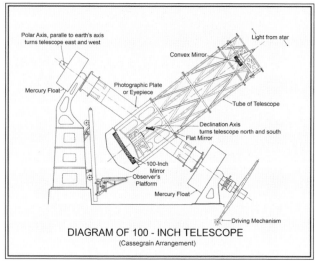

A schematic diagram shows the 100-inch telescope set to the Cassegrain focus.

Seven

SHUDDERING BEFORE
THE BEAUTIFUL

With both the 60-inch and the 100-inch telescopes up and running, the astronomy staff increased substantially. Layer by layer, this intrepid band of scientists peeled away ignorance of the universe. The laundry list of astronomy greats at Mount Wilson is a long one. Chief among them is Edwin Hubble, who would answer the long-standing question of whether or not distant spiral nebulae were part of the Milky Way or separate galaxies. His strategy was to hunt for stars in the nebula Andromeda that varied in brightness. And he found them. He found a Cepheid variable, a type of star known to be an indicator of distance. Cepheids were then found in other spiral nebulae as well, and all of them indicated that their distances were much too enormous to be part of the Milky Way. The matter was solved. Other galaxies outside of the Milky Way existed; it was not the only game in town! Mount Wilson astronomer Allan Sandage wrote, "It opened the last frontier of astronomy, and gave, for the first time, the correct conceptual value of the universe." Hubble went on to measure distances and speeds that proved the universe is expanding at an ever-increasing rate, the first concept of the big bang.

In 1919, Nobel Prize–winning physicist Albert Michelson installed the stellar interferometer he had developed for the great telescope. He was the first to make direct measurements of star sizes other than the sun. Later, he used Mount Wilson for his speed of light experiments, which stood as the most accurate measurement for 40 years.

The observational research team was highly compartmentalized and each contributed hugely: Fritz Zwicky intuited dark matter; Seth Nicholson discovered another four moons of Jupiter, the only person since Galileo to do so; Walter Baade studied different populations of stars, disclosing their formation and age; and Rudolph Minkowski studied supernovae, those gigantic explosions that can illuminate the sky for days or weeks. The great telescope advanced the world's understanding of the universe more than anyone could have dreamed.

One of the great cloudy areas puzzling astronomers was the nebula Andromeda. Nebula just means "cloudy." Was this huge cloudy mass in the Milky Way or not? The debate ran hot with astronomers taking sides and presenting their convictions at conferences, in newspapers, and in live debates. One was even called "The Great Debate," which took place at the Smithsonian Museum of Natural History to a huge crowd. Harlow Shapley of Mount Wilson firmly believed the nebula was part of the Milky Way; Heber Curtis of the Lick Observatory did not. The answer would have to wait. Here, a nine-hour exposure shows the central portion of the Andromeda nebula.

Eventually, with optics and spectroscopes performing at their highest levels, Andromeda proved to be dense with stars, as shown in this southern portion of Andromeda. This image was made following a two-hour exposure on August 24, 1925, by means of the 100-inch telescope.

Winters were often rugged on the mountaintop, sometimes dropping up to six feet of snow. For obvious reasons, car travel was then discouraged. It would be interesting to know the thoughts of the dog about this state of affairs.

The astronomers had time for fun, too. Here, they pose during a snowball fight dressed in suit and tie, the normal attire for all of the staff. Hale believed that formal dress was important in keeping a purely professional workplace in the wilderness. At work and at all meals, including breakfast, suit and tie was required. The policy continued until the late 1950s when, slowly, young scientists from Caltech managed to change the culture to more casual dress.

Astronomers came from varying backgrounds, but one thing they had in common was a rich education. All except one. A grade school dropout, Milton La Salle Humason longed for nothing more than to live on a mountaintop. The young man began his career at Mount Wilson as a mule driver carrying supplies to the observatory. He became fascinated by the work of the scientists and moved into position as janitor for the observatory in 1917 at age 26. Humason was a gambler, regularly winning poker hands off some of the astronomers. He was frequently found at the Santa Anita racetrack in the company of Adams, Baade, Minkowski, and whoever else liked to lay a fiver. Whenever the dome was opened to repair the rotation tracks, Humason asked that the telescope be positioned east. For two or three hours at the telescope, he could follow stars without any dome motion. His photographs were so superb that Adams supported his bid for a staff position. He was appointed the thankless task of scheduler for the astronomers, all of whom begged for the five darkest nights of the month. (Courtesy Caltech Archives.)

It was not just the huge mirror of the telescope that brought precision to the work of the astronomers. Enter Albert Abraham Michelson, pool shark, artist, boxer, violinist, multilinguist, renowned physicist, and Nobel Prize recipient. What he brought to Mount Wilson was his interferometer, a device that splits a beam of light in two, sends the beams along perpendicular paths, and then brings them back together. The light rays either line up perfectly or they do not. If not, the width and number of fringes produced give exquisitely delicate measurements of the rays traveling at right angles to each other. In 1920, he began using the interferometer to measure the diameter of huge Betelgeuse, one of the most luminous stars in the sky, 900 trillion miles from Earth. The measurement came in at 240 million miles diameter. The Carnegie Institution sent a telegram, "The result is staggering. The method you have added to astronomical resources is of the greatest importance." Newspapers headlined the experiment; the public and press adulation would have thrilled a rock star.

Mankind now had the ability to measure the diameter of stars with Michelson's interferometer. Here is the 20-foot instrument that was attached to the frame of the 100-inch telescope. Michelson already had a Nobel Prize, won in 1907. In fact, he and Hale had nominated each other for the prize, but as there was, and is, no prize for astronomy, the prize went to Michelson, the physicist. (Courtesy Mount Wilson Institute.)

Astronomer Francis G. Pease studies the fringes of incoming light brought in by Michelson's interferometer. The discomfort and potential danger to the astronomer is evident here at the Cassegrain focus. Pease designed much of the 100-inch telescope, taking over the project when Richey was fired. Optical manufacture and instrument design were his specialties. (Courtesy Mount Wilson Institute.)

Michelson's lifelong obsession was the speed of light. Empedocles, c. 450 BCE, believed it took time for light to travel; Aristotle did not. In 1676, Danish astronomer Ole Roemer discovered that light indeed is not instantaneous. The speed of light imposes an upper limit on how fast information can travel. Because it takes time, telescopes act as time machines, allowing astronomers to peer into the distant pass and thus see the universe as it evolved. Michelson went on to refine his experiments to measure the speed of light using Mount Wilson as one fixed point and the other point at Mount San Antonio, nearly 22 miles distant. The result was a figure that stood for over 40 years, only 4.5 miles off the current standard. A plaque was placed at the Mount Wilson Observatory site that reads, "On this pier in 1925 Albert Abraham Michelson measured the velocity of light by means of a beam of light transmitted to Mt. San Antonio and reflected back to this station." (Marvin Collins photograph; courtesy Mount Wilson Institute.)

The man with the greatest renown at the Mount Wilson Observatory, along with Hale, was Edwin Powell Hubble. He joined the faculty in 1919 and within a few years established an accurate picture of how the universe is organized, giving rise to the field of cosmology, the study of the nature of the universe. Hubble was a complicated man, an immensely talented astronomer, boxer, Rhodes Scholar, lawyer, and a master of Spanish literature. His personal relationships were as complicated as his science. His ongoing enmity with Adrian Van Maanen is especially colorful. The Monastery's mealtime seating arrangement was strict—the observer on the 100-inch sat at the head of the table. His personalized napkin ring was placed there, and others placed according to rank. Hubble had important visitors one day. He was not the observer, Van Maanen was, and Hubble was mortified that he would have to cede this honor to his rival. He arrived early, moved his napkin ring up and put Van Maanen's far down the line. He got away with it, but it did nothing to soften the bitterness between them.

The debate about nebulae raged. Is the Milky Way all there is? Hubble set out to answer the question. Were huge masses of clouds some distant conglomeration of stars, other galaxies in fact? Hubble had noticed that Milton Humason was extraordinarily gifted at taking measurements and interpreting data. He took Humason as his assistant, and the mule driver-astronomer team changed the world's understanding of the universe. Former janitor Milton Humason was awarded an honorary degree from Sweden. Here is his beautiful photograph of Cygnus Nebula.

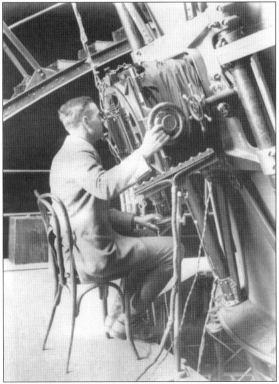

This posed photograph of Hubble at the Newtonian focus of the great telescope fails to reveal that while working he had to keep this static position for hours, in whatever temperature was ambient in the dome. In winter, it was downright frigid. (Original glass plate negative courtesy Maggie Sharma Collection.)

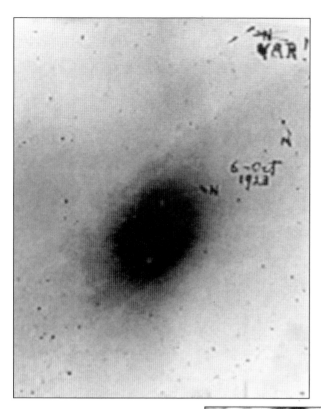

A Cepheid variable star is one whose luminosity waxes and wanes with regularity. Cepheids are large, several times the mass of the sun. Because of the physics involved, the absolute luminosity of the star can be determined. The distance is computed by comparing the absolute with the observed brightness. Hubble scouted for Cepheids, longing to "find out about the Universe." A colleague remarked, "That shows how young he is." But Hubble was on to something; he was sure of it. Here is his Eureka moment. A Cepheid in Andromeda allowed him to calculate that it was indeed a galaxy, far removed from the Milky Way. (Carnegie Institute photograph; courtesy the Huntington Library, San Marino, California.)

Hubble named his cat Copernicus. Years later, it was often acknowledged that his paradigm shift in mankind's understanding of the universe was on a par with that of his cat's namesake.

In 1913, Albert Einstein (right) had written a letter to Hale asking if a star near the sun could be photographed in enough detail to reveal the bending of light rays around its mass. Einstein's theory of relativity predicted bending, but he needed experimental proof. Hale explained in detail why that would not work, but wrote, "An eclipse method, on the contrary, appears to be very promising, as it eliminates difficulties, and the use of photography would allow a large number of stars to be measured. I therefore strongly recommend that plan." The plan was put into place with the eclipse on May 29, 1919. By year's end, Einstein's theory had brought world acclaim, and astronomers were at the center of the drama. And drama there was. The proof in May was slight. American scientists believed Einstein was wrong; the British declared, "Einstein is right." Mount Wilson would be a critical player in the outcome of the controversy. Einstein first visited Mount Wilson in 1931. Here he is at the top of the 150-foot solar tower with Walther Mayer (left) and Charles St. John (center). (Carnegie Institute photograph; courtesy the Huntington Library, San Marino, California.)

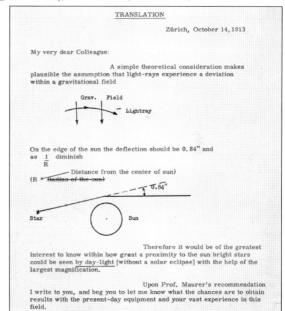

Here is the translation of Einstein's letter to Hale, courtesy of the California Institute of Technology, Einstein papers. Einstein signed it: "With all respect, yours sincerely, A. Einstein, Institute of Technology, Zurich."

Hale retired as director of the observatory in 1923 due to ill health. His successor was Walter Adams, who had come with Hale from the Yerkes Observatory. Dedicated, precise, intelligent, and an exemplary observer, Adams embodied all that Hale could desire. He was also frugal to a fault. Apparently, he demanded 25-watt light bulbs in the dome passages not because of light pollution, but to save on electricity. He mandated that the astronomer's midnight lunch be only two eggs, two pieces of bread, butter and jam, and one cup of coffee or tea. This draconian practice ended in 1946 when Adams retired and the more humane Ira Bowen took over. In the front row are, from left to right, Albert Einstein, Walter Adams, and William Campbell, director of the Lick Observatory. (Associated Press photograph; courtesy Michael Patris Collection.)

Einstein is properly impressed as he and Adams view the massive machinery of the 100-inch telescope. (Associated Press photograph; courtesy Michael Patris Collection.)

A galaxy is a huge, organized group of stars, dust, and gas held together by their mutual gravity. Earth's closest galaxy, Andromeda, is 2.5 million light-years away. Galaxies define the large-scale structure of the universe. Distances between them are immense. Einstein was impressed with Hubble's 1924 finding that there are other galaxies, but he was even more intrigued by Hubble's discovery five years later, that galaxies are rushing away from each other at enormous speed. Hubble had noticed the spectral lines from distant galaxies shifted to the red end of the spectrum. This meant that they were moving away, much as the sound of a siren changes pitch from higher when coming toward a listener and lower when moving away. It was a landmark finding. Here, Einstein views a distant star in perhaps a galaxy millions of light-years away. Hubble (center) and Walter Adams (right) watch and listen for the world's greatest scientist's comments. (Courtesy Caltech Archives.)

Einstein was certain experimental proof of his theory of relativity would come. Astronomers were sharply divided, with most believing light did not bend in the presence of the sun. A notable exception was William Campbell of Lick Observatory. Charles St. John of Mount Wilson opposed him. Immense public interest put pressure on both sides. The theoretical capabilities of the observatory had increased, but Hale admitted that Einstein's theory was too much for him and feared it would always remain beyond his grasp. As astronomers followed eclipses, the Mount Wilson astronomers discovered redshifting in the sun and in companion stars, using invaluable interferometry and spectroscopy in the pursuit. In 1925, the matter was finally resolved, and today, Einstein's theory is a part of every astronomer's tool kit. When mankind read that Hubble had discovered the universe is expanding, they learned that Einstein's theory had predicted it, but Mount Wilson astronomers had proved it. The players in this proof are, from the left to right, Milton Humason, Edwin Hubble, Charles St. John, Albert Michelson, Albert Einstein, William Campbell, and Walter Adams. They stand in front of Seymour Thomas's portrait of George Ellery Hale. (Carnegie Institute photograph; courtesy the Huntington Library, San Marino, California.)

Astronomy was not the only action on Mount Wilson. Avid car racers found the newly widened road a superb track on which to test a vehicle's performance. This impressive Gardner automobile, stripped of its fenders for lighter racing weight, was made in St. Louis, Missouri, and topped out at more than 35 miles per hour. The year is around 1925.

Not all cars on the road to the mountaintop were race cars, although that did not stop some drivers from going as fast as possible to compare the time it took to reach the summit. Visitors typically arrived at the observatory in an auto stagecoach.

Not all great discoveries were made by Hubble or Michelson. Ever since Copernicus placed the sun, not the Earth, at the center of the solar system, man sought to find himself at the center of something. His galaxy would do. It took Harlow Shapley in 1917 to displace the solar system from the center of the Milky Way by disclosing that it was actually 30,000 light-years off-center. (Courtesy Mount Wilson Institute.)

YOU ARE HERE

With the 60-inch Telescope, Harlow Shapley Found Our Place in the Milky Way Galaxy -- 1918

MOUNT WILSON
OBSERVATORY

Astronomer Walter Baade worked at Mount Wilson Observatory from 1931 to 1958. He discovered that there were two types of Cepheid variable stars, revealing that the size of the universe was double the size calculated by Hubble in 1929. He also spotted that the Crab Nebula was a remnant of the 1054 supernova, a gigantic exploding star that lights up the sky for days or weeks and spews heavy elements throughout. "We are made of star stuff," as Carl Sagan famously once said.

The 100-inch telescope captured the moon in exquisite detail. This is the northern portion of the moon 18 days after the sliver of a crescent of the waning moon. This image dates from August 7, 1925.

Due to the tilt of Earth's axis, many more stars are visible to the naked eye in the southern hemisphere. The same amount exists in the northern hemisphere, but they are not discernible. Eventually, though, astronomers were able to photograph them. This picture of the southern hemisphere shows the Southern Cross on the left half of the image. Blue stars are young, hot bright stars, red ones are cooler and older. The density of stars in the southern sky is breathtaking, a shuddering before the beautiful. This image was created with a three-inch Ross-Tessar camera.

Eight

CHANGING TIMES

Hale retired from the Mount Wilson Observatory in 1923 due to ill health but remained in charge of policy and research. His friend Henry Huntington had bequeathed him land on Halliday Road, not far from Caltech, where he built a small solar laboratory to continue his studies of the sun. Distinguished astronomers, including Einstein, came to marvel at his images of solar prominences reaching thousands of feet into the coronasphere. Hale had refined his spectrohelioscope, enabling it to capture the entire hydrogen atmosphere of the sun for the first time in history. Always happy to call himself an amateur, Hale reveled in his solitary work at the little observatory, a respite from his formerly hectic life. The entrance to the Hale Solar Laboratory library was embellished with a bas-relief tribute to Ikhnaton, depicted there as the sun's rays converging to hands that grasp the symbol of life itself. It was a copy from a Theban tomb, a nod to Hale's lifelong passion for Egyptology.

It is a truism that times change and instruments that were once state of the art are replaced by more powerful ones. Light pollution had endangered astronomy on Mount Wilson; new technology was needed. It came in the form of a brilliant design called adaptive optics, a computer-driven deformable mirror system that improved clarity by greatly reducing aberrations caused by heat and light pollution. In addition, a program for schoolchildren to remotely access the 24-inch telescope from their classrooms was developed. Schools in India, Australia, South America, Canada, and the United States participated.

Meanwhile, Hale had his heart set on developing a massive 200-inch telescope. This he did, and Palomar Observatory came into being at the top of Palomar Mountain, northeast of San Diego. Hale died in 1936 and did not live to see his beloved telescope begin its work in 1948, but it was his design and far-reaching vision that made the huge instrument possible. Fittingly, it was named the Hale telescope.

This beautiful library was a sanctuary for Hale in his retirement. In a separate building from the telescope dome, it had a machine shop in the basement and all manner of research materials in the library. An ancient Persian astrolabe, an armillary sphere, and Egyptian paintings graced the room.

The bas-relief by Lee Lawrie over the entryway to the library contains much iconic Egyptian art. The sun's rays converge at the top into hands reaching out to grasp life-giving heat and light. The side panels depict various scenes in the times of Akhenaten, the 10th pharaoh of Egypt, who worshipped Aten, the disk of the sun. He was the father of King Tut (Tutankhamen) and the husband of Nefertiti. (Carnegie Institute photograph; courtesy the Huntington Library, San Marino, California.)

Einstein visited the Hale Solar Laboratory several times, always exulting in discussions of the composition of the sun and therefore other stars more distant. Here, he is pictured giving a talk in the Hale Library at the Carnegie Institute in Pasadena. (Carnegie Institute photograph; courtesy the Huntington Library, San Marino, California.)

This modern-day photograph highlights the Hale Solar Laboratory dome. The library's tile roof is just visible on the left. A plaque erected in 1989 sits on the pillar at the end of the driveway. The property was sold by the Mount Wilson Institute and is now privately owned and has ceased to function as a center for astronomy.

The plaque establishes the Hale Solar Laboratory as a national historic landmark by the National Park Service, United States, Department of the Interior. It reads: "This property possesses national significance in commemorating the history of the United States of America."

Astronomers Seth Nicholson and Edwin Pettit are checking the mechanics of their motion picture camera that will make a "movie" of the solar eclipse to take place in the California desert on April 28, 1930. Nicholson headed the expedition. Unfortunately, clouds covered the sun at the moment of totality, and the hoped-for photometry of the corona and limb was dashed. (ACME Press photograph; courtesy Michael Patris Collection.)

In 1934, the Angeles Crest Highway to Redbox Junction was completed. A dirt road for a few more miles completed the journey to the observatory, and the road now gave cars full access to the mountaintop. The Mount Wilson Hotel continued to do a brisk business and later even had a swimming pool for a time.

While astronomical findings continued at Mount Wilson and the Hale Solar Laboratory, Hale forged ahead with his dream of a 200-inch telescope. A honeycomb structure was devised into which the glass would be poured. The support of the honeycomb would keep the 17-foot glass from sagging. The grinding and polishing were kept true via lightwave measurements, as no mechanical device could measure the 1/1 millionths of an inch leeway allowed for the parabola. The telescope was five stories tall and would weigh in at 530 tons. An innovation was the mount empowering the telescope to be pointed from the North Pole right down to the southern horizon. The design was ready, and a rough grinding of the mirror was completed by 1940, but the Second World War delayed construction. In 1948, the immense apparatus viewed the heavens as never seen before, able to peer into space a distance of 400 million light-years.

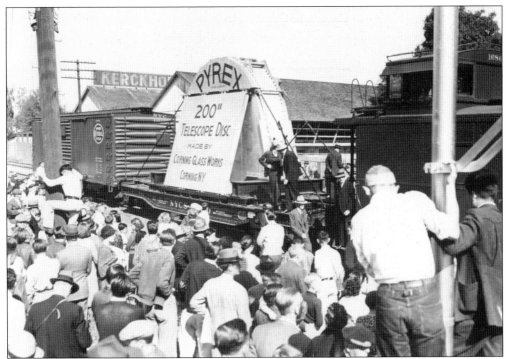

Corning Glassworks in New York was chosen to pour the glass for the mirror. As with the 100-inch disk, there were problems. A borosilicate glass had been chosen, and while it turned out to be a perfect choice, it had to be poured a few times before it was deemed satisfactory. Here, a crowd of curious spectators watch as the massive 200-inch disc is delivered by freight train to the Santa Fe Railroad's Lamanda Park Station in East Pasadena. It would be taken to Caltech for grinding and polishing. (Associated Press photograph; courtesy Michael Patris Collection.)

The innovative honeycomb structure met expectations, keeping the huge glass from sagging. This was the first use of the design. Borosilicate is Pyrex, the same material that is in old household glass measuring cups and bakeware. It has a low thermal expansion rate and does not easily crack. After 1992, the new measuring cups, still called Pyrex, were changed to a soda-lime formula with a faster expansion rate that can shatter in household microwave ovens.

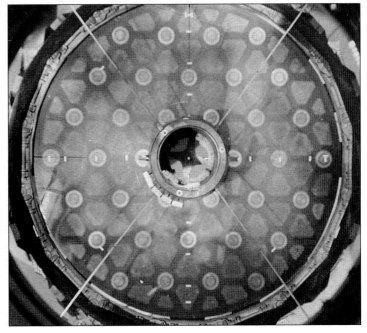

The magnificent telescope is safely ensconced in its pristine dome. Visual observation was by now a thing of the past. Photographic plates are infinitely more sensitive to light than the human eye, and time exposures allow astronomers to obtain pictures of stars far too faint for the human eye to see. (ACME Newspictures photograph; courtesy Michael Patris Collection.)

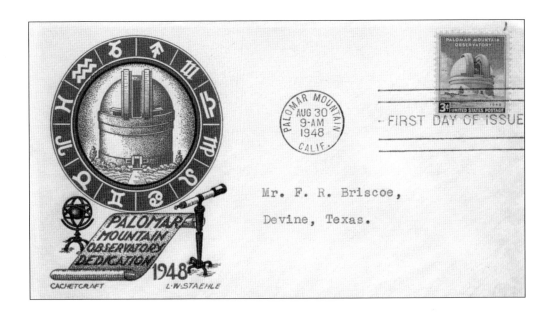

On August 30, 1948, the first postage stamp of the great Hale telescope was issued. A postcard stamp cost 3¢. During the first three quarters of the 20th century, Hale had built the four largest telescopes in the world, from Yerkes to Palomar. A preponderance of advances in astrophysical research was done on one or another of Hale's telescopes.

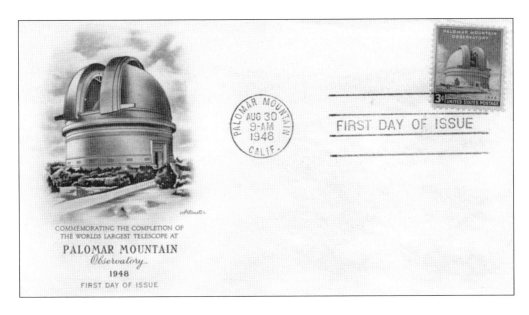

The entrance to Skyline Park captured in this Mirro-Krome postcard was photographed by Hubert A. Lowman for the H.S. Crocker Company of Los Angeles in the early 1960s. The information provided by the publisher on the back of the card states, "Mount Wilson Skyline Park, California. Entrance gate to the 720 acre mile high park."

The Mount Wilson Institute was formed and began operating the Hooker telescope in 1989 after it had closed four years previously. The Carnegie Institute had abandoned support of Mount Wilson in favor of astronomy centers in Chile, but it was not long before a new technology came along. Adaptive optics, first developed for the military and adapted by Chris Shelton, shown here, for the Cassegrain focus of the 100-inch telescope, changed the playing field. (Beatty photograph; courtesy Mount Wilson Institute.)

The results of the adaptive optics system were immediate. A computer-controlled deformable mirror compensated for distortion caused by atmospheric and light pollution. The result is image clarity comparable to that of space telescopes. Here, a triple star, Zeta Cancri, is imaged without adaptive optics. Only two stars are visible. The stars' images blur together by passage through the atmosphere. (Carnegie Institute photograph; courtesy the Huntington Library, San Marino, California.)

Using adaptive optics, three light images of the stars are clearly visible. The light that had been spread has been gathered together to form three narrow spikes, and a triple star emerges. This technology put the 100-inch telescope back in business. (Carnegie Institute photograph; courtesy the Huntington Library, San Marino, California.)

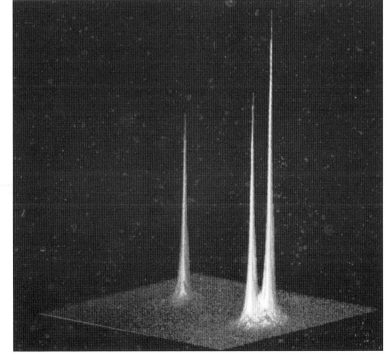

Another new technology was afoot at the observatory. In 1993, the Mount Wilson Institute sponsored a program to bring astronomy to the classroom. Jet Propulsion Laboratory (JPL) engineer Gilbert Clark directed the very successful program called Telescopes in Education (TIE). The project allowed schoolchildren across the globe to remotely access and control a telescope on Mount Wilson and take astronomical photographs in real time. The 24-inch telescope, languishing in a basement at Caltech, had been used in the 1960s to image the moon to determine whether or not the lander would have a solid surface or sink into thick dust. Clark revived the telescope with software for his program, drawing in grade school students from Canada, America, India, Chile, Australia, England, and parts of Europe. (Courtesy Caltech, NASA, JPL.)

Here, Gilbert Clark is in Queensland, Australia, describing to middle school students the celestial objects they have imaged on their computer using the Mount Wilson telescope. (Courtesy NASA, JPL.)

This image of the Whirlpool Galaxy with a supernova at its core was taken by elementary school students using the 24-inch telescope. It is a beautiful rendition of a spiral galaxy showing arms that are dust lanes of star formation. Whirlpool is 28 million light-years from the Milky Way galaxy and about 40 percent of its size. (Courtesy NASA, JPL.)

The amazing Horsehead Nebula was captured by Native American middle school students. It is dark because dust and hydrogen block the stars behind it. The nebula is part of the Milky Way in the easternmost portion of Orion's belt, 1,500 light-years from Earth. It is visible from Earth. The program brought students into astronomy, some of whom went on to become astronomers themselves, a well-deserved tribute to Gilbert Clark. (Courtesy NASA, JPL.)

Nine

100 Years On

Completion of Angeles Crest Highway in 1934, wide enough for any type of vehicle or any weight, meant that until the turn of the millennium, transportation ceased to present difficulties. When Georgia State University agreed to build a large interferometry system on the observatory grounds, equipment and materials once again rolled up the mountain. This time it was easy. That is, until the telescopes had to be placed. The Center for High Angular Resolution Astronomy (CHARA) would have a series of six one-meter telescopes contained in small dome enclosures set in a Y-shaped pattern. The resolution was the sum of the baseline distance among the telescopes, in this case 350 meters, a resolution equal to a thousand-foot mirror, or the angular size of a nickel seen from a distance of 10,000 miles. Binary and triple stars, distances, diameters, masses, and luminosities were clearly defined; it was unparalleled for the study of stellar nurseries. The Mount Wilson site was chosen by director Harold McAlister because of its stable air and number of clear days per year.

Figuring out how to get the six hefty domes to the site in the shortest time at the least cost was the quandary. Sea West Enterprises won the contract for designing and engineering the components as well as fabricating and installing the domes for the telescopes. They decided to use a helicopter to place the domes from where they were assembled in the parking lot on the peak to their enclosures on the site, avoiding any impact on local terrain from widening trails or staging of large cranes and other high-lift equipment. Hale's lifelong quest to unravel the mystery of stellar evolution was advancing by leaps and bounds.

In his lifetime, Hale's visionary accomplishments dominated the world of astronomy. Among other accomplishments, he succeeded in establishing the California Institute of Technology as a world-class institution; ensured that the Henry E. Huntington Library, Gardens, and Art Gallery remained a center of academic distinction; and developed the international strength of the National Academy of Science. George Ellery Hale was truly sui generis.

Groundbreaking began in 1996 for the world's largest optical telescope array complex, four years in the making at the modest price tag of $15 million. The director of the Center for High Angular Resolution Astronomy (CHARA), Dr. Harold McAlister of Georgia State University, is pictured on the right in this dedication ceremony photograph.

Harpist Mary Cragg, wife of Gilbert Clark of Telescopes in Education, is ready to bring her instrument to the platform of the 100-inch telescope. She provided ethereal music for the dedication ceremony of CHARA, much to the delight of the crowd attending one of the largest events held at the Mount Wilson Observatory.

After mapping the position of the six domes in order to maximize the area of sky that could be seen by each, the crew poured cement to form the base of each 32-foot-tall enclosure dome. Here, the completed domes stand on the summit. (Courtesy George Elder.)

A dome is shown both closed and open. In the open dome, the interferometer is visible. The meter-wide mirrors for each dome were fabricated in Russia. McAlister's brainchild of a large interferometry array was about to bring groundbreaking astronomy back to Mount Wilson. (Courtesy Mount Wilson Institute.)

When Sea West Enterprises Inc. was awarded the contract to work with Georgia State University in 1997, it agreed to install six telescope enclosures each weighing 7.5 tons onto their pier locations, creating "the most powerful instrument of its kind in the world," according to CHARA director Hal McAlister. On January 24, 1999, the crews assembled before daylight for the spectacular airlift. Sea West CEO Jack Simison was grateful for the cold, dense air, which would hasten the work ahead. A rear-facing pilot was able to visually guide the helicopter to its exact placement, but due to weight, the chopper carried only 10 minutes worth of fuel. Simison figured the transaction to deliver and install each enclosure would take seven minutes, including refueling. When he advised McAlister that the operation would be completed in an hour, McAlister thought it implausible. A crew of 30 worked in stunning synchrony; all enclosures were installed in 54 minutes. McAlister's original aerial photograph of the helicopter could not be obtained for this work; shown here is a similar sky crane type lifting a Howitzer cannon for artillery use. (Courtesy Michael Patris Collection.)

Risk and pain were present. A helicopter's rotors with trick line attached will generate a field, and the first person to touch the line experiences a nice low-amp 110-volt hit. The crew democratically took turns. Shown here is the dome of one of the telescopes. Each of the mirrors reflects its light via a vacuum tube (lower left) leading to the central beam-combining facility, where the light waves of each of the telescopes are brought into phase with one another. (Photograph by and courtesy of Steve Padilla.)

The Beam Synthesis Facility is a stadium-length building where the light rays are bounced between mirrors until all are cued up to meet at one location. This clever technology simulates the properties of a single large instrument by combining the light from several smaller ones. It is an inexpensive way to build a gargantuan mirror.

Fritz Zwicky, a longtime faculty member of Caltech, was an outstanding observational astronomer at Mount Wilson and, later, Palomar. He was a trailblazer, developing original theories that have had a momentous influence on the world of astronomy. His work spanned several decades in the studies of supernovae, neutron stars, cosmic rays, and, in 1933, his landmark hypothesis of the existence of dark matter. (Courtesy Caltech Archives.)

Were there women in astronomy in the early 20th century? Very few. Maria Mitchell at Vassar was one; Henrietta Leavitt at Harvard Observatory was not an astronomer but a "computer," the title given to women scrutinizing photographs and performing mathematical calculations. Leavitt was gifted. She was also profoundly deaf. She examined tens of thousands of plates several times, looking for a star that varied in luminosity, and became an expert at Cepheid variable stars, which Hubble would use to calculate the distance to Andromeda. Mount Wilson had its own "computers," notably Louise Ware, brought by Hale from Yerkes to Mount Wilson in 1906 where she remained until 1942. Women did not work at the mountaintop but at the observatory's offices in Pasadena. Ware painstakingly recorded positions of dark filaments on the disc of the sun and coauthored several scientific articles with astronomer Charles St. John. At Harvard, Annie (Jump) Cannon cataloged the classification of more than 350,000 stars according to temperature; Cecilia Payne-Goposchkin was the first woman to receive a doctorate in astronomy, in 1925. This humorous drawing slyly points the acumen of women in astronomy.

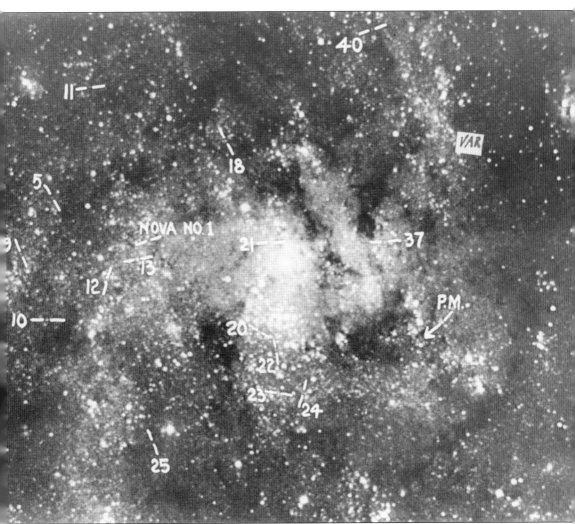

Technology marches on. State-of-the-art advances are superseded by further refinements and developments. The magnificent 100-inch telescope, long the greatest in the world, was eventually replaced by Palomar and then by large interferometry systems. Progress may look like a steady arrow forward, but it is not. Setbacks and errors characterize all human endeavors; an example is the photograph above. Milton Humason had shown astronomer Harlow Shapley a photograph of stars that appeared in some plates but not in others. He had calculated their distances and concluded they had to be far outside the Milky Way galaxy. Shapley looked at Humason's images and said it was not possible as the Andromeda Nebula was simply a nearby gaseous cloud within the Milky Way, whereupon Humason wiped the photographic plate clean of his identifying marks. This incident took place a couple of years before Hubble discovered Andromeda's Cepheid variable star. If Shapley had taken Humason seriously, would Humason now be lauded for discovering galaxies beyond the Milky Way? Very likely, but due to Shapley's human error, or hubris, it was not to be.

HOTEL   MT. WILSON, CALIFORNIA.

Changes were afoot. In 1963, Metromedia built a pavilion and installed an animal zoo for children at the Mount Wilson Lodge, formerly the Mount Wilson Hotel. It became a popular destination for families seeking a respite from the city. Food was available, and the hiking trails introduced many young people to the wonders of spending time in nature. Some even ventured to explore the observatories, and there is more than one story of a youngster who caught the thrill of exploring the heavens from these early experiences on the mountain.

Thirty years later, Skyline Park looked like this. The zoo is no more; people routinely visited for the chance to breathe clean air at a time when Los Angeles suffered from considerable pollution.

The Carnegie Institute had pulled funding from the Mount Wilson Observatory in 1985, favoring the darker skies of Chile. The 100-inch telescope was mothballed, but amateur astronomer Robert Ferguson and Dr. Arthur Vaughn of the Jet Propulsion Laboratory in Pasadena spearheaded a drive to form the Mount Wilson Institute with an illustrious board of trustees. A permanent agreement transferred the management of the observatory to the Mount Wilson Institute for an indefinite period of time, guaranteeing stability into the foreseeable future. In 1991, Robert Jastrow became director and board chairman of the institute, and Dr. Sallie Baliunas became deputy director. Jastrow, astronomer and planetary physicist, was a well-known author and populizer of astronomy, hosting over 100 CBS-TV network programs on space science. He also served as director of NASA's Goddard Institute for Space Studies from 1961 to 1981. Jastrow's commitment to Mount Wilson's continued development was the bedrock for advances in adaptive optics and infrared astronomy and was instrumental in supporting the decision to add the CHARA array to the observatory. (Courtesy Michael Patris Collection.)

Interferometry became a major tool in the exploration of the universe. Michelson's majestic 20-foot interferometer was used on the 100-inch telescope throughout the 1920s to measure the distances between close binary stars and to make the first-ever measurements of the angular diameters of stars. Currently, the historical piece seen here is housed in the CHARA Exhibit Hall. (Courtesy Mount Wilson Institute.)

In 1998, Nobelist Dr. Charles Townes of the University of California, Berkeley began the operation of an infrared spatial interferometer, first consisting of two 1.6-meter telescopes placed 30 feet apart. Because it operates in the mid-infrared portion of the spectrum, the interferometer cuts through gas and dust allowing astronomers to study in great detail stars in the process of forming, stars hidden in the shadow of brighter stars, and Cepheid variables changing in luminosity and size. (Courtesy Mount Wilson Institute.)

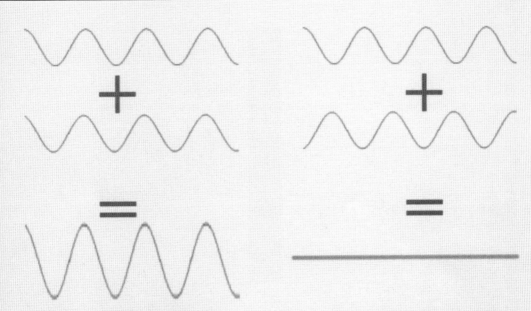

Constructive interference

Destructive interference

On the left, the waves are in phase, that is, the troughs and peaks align producing a wave that is twice as big but otherwise the same as the two original waves. On the right, the waves are of out phase, the peaks of one coinciding with the troughs of the other. Adding these waves together results in no wave at all. They cancel each other out. The phase difference between the two beams creates a pattern of light and dark areas called interference fringes. In light areas, the beams have added together to become brighter. In the dark areas, beams have subtracted from each other. It is a precise way of measuring with extraordinary accuracy. In astronomy, interferometers are used for mapping the features on the surface of stars, measuring the diameters of stars, resolving distances between close binary stars, and a host of other applications. Recently, laser interferometry was used to discover gravitational waves, predicted by Einstein more than a century ago, but it was the Mount Wilson Observatory that promoted the interferometer as a precise and detailed scientific instrument. (Courtesy www.explainthatstuff.com.)

Earl C. Anthony
Television Transmitter
W-21
Mt. Wilson, Calif. Elev. 6000 ft

It was 1947 when the first transmitting antennae went up on Mount Wilson, a half mile from the observatory. The pioneer was a new television station in Los Angeles—KTLA. As the number of stations grew, Southern Californians were able to receive broadcasts of both radio and television from a variety of stations.

The tallest antenna in the group is a 972-foot spike owned by KCBS-TV. In the early 1990s, director Robert Jastrow became concerned that signals from the stars would soon be drowned out by radio and television broadcasts from the increase in transmission towers. Most research would be curtailed or eliminated. Southern California Site Facilities Inc. agreed to limit the power of signals aimed toward the observatory to less than 300,000 watts and to shield tower lights from the observatory. Astronomers breathed a sigh of relief.

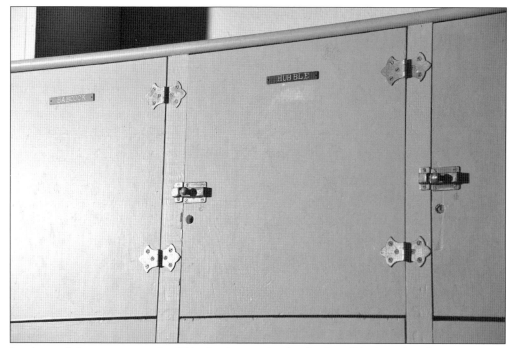

Just beyond the entryway to the 60-inch telescope is a bank of small lockers that the astronomers used to store belongings while they were observing. Pictured here is Edwin Hubble's locker, his name still visible on the metal tape. (Courtesy George Elder.)

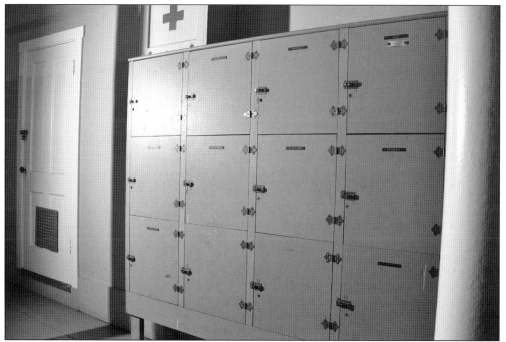

Other famous Mount Wilson astronomer's names remain in this section of the lockers. The names are, from left to right, as follows: (top row) unidentified, Babcock, Hubble, and Joy; (second row) unidentified, O. Wilson, R. Wilson, and Zwicky. (Courtesy George Elder.)

One hundred and seventeen years later, the Littrow spectroscope stands in the Snow telescope as it did in 1905. After all these years, it still performs perfectly, serving as a teaching tool for undergraduate astronomy majors and summer internship programs.

The observatory looked very much like this even 100 years ago. The original workhorses, Hale's five telescopes, have not changed over the century. CHARA, a computer building at the 150-foot tower, and Berkeley's special infrared interferometer are among the few additions. It is hard to believe the contrast in American life from 1904 to 2004. In 1904, the average life expectancy was 47 years, 14 percent of homes had a bathtub, and only 8,000 cars and 144 miles of roads existed in this nation. California was the 21st most populous state, and the average wage was 22¢ per hour. Nearly all births took place at home, and influenza, pneumonia, tuberculosis, and diarrhea were leading causes of death. Las Vegas, Nevada, boasted a population of 30; 20 percent of adults could not read or write; and only six percent of all Americans had graduated from high school. (Courtesy Mount Wilson Institute.)

One of the few women in 1904 to complete high school, this young woman, Lena B. Johnson of Brooklyn High School in New York, stands dignified and proud of her achievement.

For nearly four decades, the Mount Wilson Observatory was home to the world's two largest telescopes. The phenomenal output of research changed forever man's view of the universe. It all started with Hale's dream of what could be. Here, in the earliest days on the mountain, Hale and Ferdinand Ellerman make plans to establish a center of science at the summit. "The art of research makes difficult problems soluble by devising means of getting at them," observes Nobel Prize winner Peter Medawar in *Pluto's Republic*. Astrophysics may seem remote from everyday life, but what could be more intriguing than understanding how the universe is organized, where its margins are, what it consists of, what existed before the solar system, how it was formed, how it will end, and if there is other sentient life? These were the large questions that Hale set out to answer at the turn of the 20th century.

Establishing the Mount Wilson Observatory took money as well as vision. In this early photograph, Hale and Andrew Carnegie walk, intently discussing the funding of the fledgling observatory. Without the generous support of the Carnegie Institute over many years, Hale's dreams would not have materialized. The intersection of search, discovery, and interpretation is a triad beguiling to all and wherever it occurs, powerful instruments and methods will be the means. Constructing, transporting, and placing these marvels will be the obstacles. Advances in 21st-century astronomy astound mankind, as space-based telescopes photograph the early universe. For most people, it is new science; to the heavens, it is history. Though often deferred, Hale's visions triumphed in the end. Helen Wright's authoritative biography *Explorer of the Universe: A Biography of George Ellery Hale* took its title from a quotation on the central dome of the National Academy of Science in Washington, DC: "To science, pilot of industry, conqueror of disease, multiplier of the harvest, explorer of the universe, revealer of nature's laws, eternal guide to truth." Indeed, astronomer George Ellery Hale was the 20th-century's pioneer explorer of the universe. (Carnegie Institute photograph; courtesy the Huntington Library, San Marino, California.)

BIBLIOGRAPHY

Clerke, Agnes M. *Problems in Astrophysics*. London: Adam & Charles Black, 1903.

Crelinsten, Jeffrey, *Einstein's Jury: The Race to Test Relativity*. Princeton, NJ: Princeton University Press, 2006.

Gribbin, John. *Companion to the Cosmos*. New York: Little, Brown and Company, 1996.

Hale, George E. *Ten Year's Work of a Mountain Observatory: A brief account of the Mount Wilson solar observatory of the Carnegie Institution of Washington*. Washington, DC: Carnegie Institute of Washington, 1915.

Jastrow, Robert. *Red Giants and White Dwarfs*. New York: W.W. Norton & Company, Inc., 1990.

Johnson, George. *Miss Leavitt's Stars: The Untold Story of the Woman Who Discovered How to Measure the Universe*. New York: W.W. Norton & Company Inc., 2005.

McAlister, Harold A. *Seeing the Unseen: Mount Wilson's role in high angular resolution astronomy*. Bristol, England: IOP Publishing Ltd, 2020.

Overbye, Dennis. *Lonely Hearts of the Cosmos: The Story of the Scientific Quest for the Secret of the Universe*. New York: Little, Brown and Company, 1996.

Rees, Martin. *Before the Beginning: Our Universe and Others*: Reading, MA: Helix Books, 1997.

Robinson, John W. *The Old Mount Wilson Trail*. City of Industry, CA: Big Santa Anita Historical Society, 2001.

Sandage, Allan, R. *Centennial History of the Carnegie Institution of Washington Vol. 1: The Mount Wilson Observatory*. Cambridge: Cambridge University Press, 2004.

Sharma, Maggie. *From Footsteps to Flying Machines*. The Westerners, Los Angeles Corral, Brand Book 22. Spokane, WA: Arthur H. Clark Company, 2004.

———. *The Fugitive Constant: Michelson and the Speed of Light*. Los Angeles, CA: The Branding Iron, Los Angeles Westerners Corral, 2005.

Shirley, Christine. *The Hale Solar Laboratory–740 Halladay Road, Pasadena–A Guided Tour*. Pasadena, CA: Christine Shirley, 1985.

Wright, Helen. *Explorer of the Universe: A Biography of George Ellery Hale*. New York: E.P. Dutton & Co. Inc., 1966.

About the Organizations

The Mount Wilson Observatory

The most scientifically productive astronomical observatory in history, this was the preeminent facility in the world in both stellar and solar studies during the first half of the 20th century. Modern instrumentation has enabled both the original superb telescopes and the more recently built facilities to continue Mount Wilson's pioneering heritage in the new fields of study. Now run by the Mount Wilson Institute, its mission is to manage and promote the Mount Wilson Observatory for current scientific research, historic preservation, education, and public engagement (www.mtwilson.edu).

The Mount Lowe Preservation Society

The Mount Lowe Preservation Society Inc. (MLPSI, www.mountlowe.org) is a California nonprofit 501(c)3 educational foundation formed by a group of dedicated individuals committed to opening a permanent museum in Pasadena, California. Our building houses and will display memorabilia, artifacts, photography, ephemera, and literary works for the purpose of educating the public about Southern California's transportation evolution. Our collections include extensive materials from the Mount Lowe Railway and its founder—builder and engineer Thaddeus S.C. Lowe and David J. Macpherson—as well as successor companies, including the Pacific Electric Railway and the Southern Pacific Railway. With a focus on railroad and trolley travel, our collections also include an indoor railroad signal garden, antique automobiles, transportation photography, display models, Civil War ballooning materials from the late Thaddeus S.C. Lowe, and an extensive railroad-related research library.

Discover Thousands of Local History Books
Featuring Millions of Vintage Images

Arcadia Publishing, the leading local history publisher in the United States, is committed to making history accessible and meaningful through publishing books that celebrate and preserve the heritage of America's people and places.

Find more books like this at
www.arcadiapublishing.com

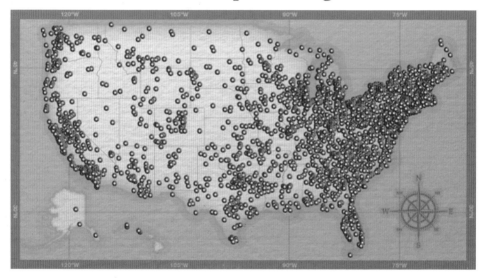

Search for your hometown history, your old stomping grounds, and even your favorite sports team.

Consistent with our mission to preserve history on a local level, this book was printed in South Carolina on American-made paper and manufactured entirely in the United States. Products carrying the accredited Forest Stewardship Council (FSC) label are printed on 100 percent FSC-certified paper.

MADE IN THE USA